T0269230

NEW CONCEPTS IN THE MANAGEMENT OF SEPTIC PERIANAL CONDITIONS

NEW CONCEPTS IN THE MANAGEMENT OF SEPTIC PERIANAL CONDITIONS

RIYADH MOHAMMAD HASAN
Assistant Professor (MB. ChB., C.A.B.S., General Surgeon),
Department of Surgery, Al-Kindy Medical College,
University of Baghdad, Baghdad, Iraq

BATOOL MUTAR MAHDI
Professor (MB. ChB., MSc., FICMS-Path., Clinical Immunology),
Head of HLA research Unit,
Department of Microbiology, Al-Kindy Medical College,
University of Baghdad, Baghdad, Iraq

ACADEMIC PRESS

An imprint of Elsevier

Academic Press is an imprint of Elsevier
125 London Wall, London EC2Y 5AS, United Kingdom
525 B Street, Suite 1650, San Diego, CA 92101, United States
50 Hampshire Street, 5th Floor, Cambridge, MA 02139, United States
The Boulevard, Langford Lane, Kidlington, Oxford OX5 1GB, United Kingdom

Notices
Knowledge and best practice in this field are constantly changing. As new research and experience
broaden our understanding, changes in research methods, professional practices, or medical treatment
may become necessary.

Practitioners and researchers must always rely on their own experience and knowledge in evaluating
and using any information, methods, compounds, or experiments described herein. In using such
information or methods they should be mindful of their own safety and the safety of others, including
parties for whom they have a professional responsibility.

To the fullest extent of the law, neither the Publisher nor the authors, contributors, or editors, assume
any liability for any injury and/or damage to persons or property as a matter of products liability,
negligence or otherwise, or from any use or operation of any methods, products, instructions,
or ideas contained in the material herein.

Library of Congress Cataloging-in-Publication Data
A catalog record for this book is available from the Library of Congress

British Library Cataloguing-in-Publication Data
A catalogue record for this book is available from the British Library

ISBN 978-0-12-816111-1

For information on all Academic Press publications
visit our website at https://www.elsevier.com/books-and-journals

Working together
to grow libraries in
developing countries

www.elsevier.com • www.bookaid.org

Publisher: Mica Haley
Senior Acquisitions Editor: Stacy Masucci
Editorial Project Manager: Megan Ashdown
Production Project Manager: Punithavathy Govindaradjane
Cover Designer: Christian Bilbow

Typeset by SPi Global, India

DEDICATION

I would like to dedicate this book to:
1. My beloved mother for her kindness, endless support, and loving spirit that endure and sustain me.
2. The memory of my darling father, a smart man whom I still miss every day.
3. My adored brothers, Dr. Adil and Dr. Ali Ghalib.
4. My wonderful nephew, Sajad.
5. Our students, past, present, and future.

Batool Mutar Mahdi

CONTENTS

PREFACE

This book is a concise review of medically important aspects of a surgical problem. It covers both basic and clinical aspects of perianal diseases in adults. Its two major aims are to assist students who prepare for specialty in colorectal diseases and its management, and to help people who want to understand more about this condition with brief and flexible sources of information. This book presents current, medically important information in the rapidly changing field of anorectal diseases. It includes updated information on such topics as perianal abscess and fistula. Our goal is to provide readers with an accurate source of clinically relevant information at different levels of medical education. These aims are achieved by utilizing several different formats, which should make the book useful to students and readers with varying study objectives and learning styles. The information is presented succinctly with stress on making it clear, interesting, and up to date.

We believe that readers appreciate a book that presents the essential information in a readable and interesting format. We hope you find this book meets those criteria.

Riyadh Mohammad Hasan

Batool Mutar Mahdi

ACKNOWLEDGMENTS

We are indebted and thankful to our God for helping us to complete this book and in making this book a reality.

INTRODUCTION

Batool Mutar Mahdi

Anorectal diseases are challenging conditions for both patients and surgeons. The history of treatment of these conditions can be traced back to ancient times. The acute stage is anal abscess, which originates from an infection arising in the cryptoglandular epithelia lining the anal canal. The internal anal sphincter is believed to serve normally as a barrier to infection passing from the gut lumen to the deep perirectal tissues. This barrier is breached by the crypts of Morgagni, which can penetrate through the internal sphincter into the intersphincteric space. Once infection gains access to the intersphincteric space, it has an easy access to the adjacent perirectal spaces. The severity and depth of the abscess are quite variable, and the abscess cavity is often associated with formation of a fistulous tract. Fistula-in-ano is a chronic challenging pathology for which treatment attempts have been traced back to the days of Hippocrates (Fig. 1), who inserted horse hair with lint into the fistula, which was periodically tightened.

Later on, the Middle Eastern physician Albucasais (AD 936–1013) (Fig. 2), the medieval physician John of Ardene (AD 1307–92), and, in 1835, Fredrick Salmon (Fig. 3) practiced a variety of methods to treat fistulae-in-ano.

Fig. 1 Hippocrates (370–460 BC).

Fig. 2 The Middle Eastern physician Abu al-Qasim Khalaf ibn al-Abbas Ali Known in the west as Albucasais (AD 936–1013).

Fig. 3 Fredrick Salmon (AD 1796–1868).

This book will try to shed a light on the most recent and practical methods in the treatment of perianal diseases with less cost and fewer settlements in the hospital.

CHAPTER 1

Anatomy of the Anal Region

Riyadh Mohammad Hasan

Contents

1 INTRODUCTION

The anal canal is the terminal part of the large intestine. It is situated between the rectum and anus, below the level of the pelvic diaphragm. It lies between the right and left ischiorectal fossa. The anal canal consists of three parts. The zona columnaris is the upper half of the anal canal and is lined by simple columnar epithelium, and below the pectinate line is divided into two zones, separated by Hilton's white line (zona hemorrhagica and zona cutanea) and lined by stratified squamous nonkeratinized and stratified squamous keratinized epithelia, respectively. In humans, it is about 2.5–4 cm long, extending from the anorectal junction to the anus. It is directed downwards and backwards, and is surrounded by inner involuntary and outer voluntary sphincters which keep the lumen closed in the form of an anteroposterior slit (Dobson et al., 2003).

2 EMBRYOLOGY

The anal canal represents the site of junction of the hindgut (endoderm) with the proctodeum. The proctodeum is an invagination of the skin (ectoderm) that joins the hindgut, and forms the site of the mid-anal canal. The hindgut follows the midgut in the embryo and extends from the posterior intestinal

New Concepts in the Management of Septic Perianal Conditions
https://doi.org/10.1016/B978-0-12-816111-1.00001-X

portal to the cloacal membrane (Fig. 1). It gives the left one-third to one-half or distal portion of the transverse colon, the descending colon, the sigmoid colon, the rectum, the upper portion of the anal canal, and part of the urogenital system. The terminal part of the hindgut enters into the cloaca, which is an entoderm-lined cavity that is in direct contact with the surface ectoderm. The cloacal membrane is composed of entoderm of the cloaca and ectoderm of the proctodeum or anal pit, and the terminal part of the hindgut, the cloaca, receives the allantois ventrally and the mesonephric ducts laterally. During development, the urorectal septum forms in the angle between the allantois and the hindgut. As it grows caudally toward the

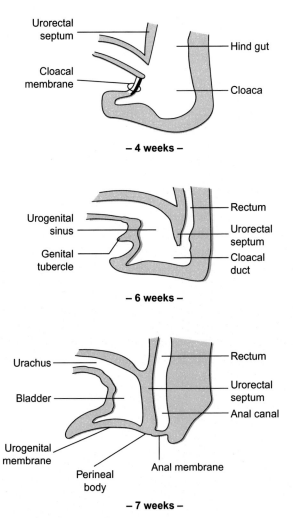

Fig. 1 Embryology of the anal canal.

cloacal membrane, it divides the cloaca into an anterior portion, the primitive urogenital sinus, and a posterior part, the anorectal canal. By 7 weeks of age, the urorectal septum reaches the cloacal membrane and fuses with it. Thus, the membrane is divided into a posterior anal membrane and a larger anterior urogenital membrane. The area of fusion of the urorectal septum and the cloacal membrane becomes the primitive perineum or perineal body. In week 9, proliferation of mesenchyme around the anal membrane raises the surrounding ectoderm to form a shallow pit, the anal pit or proctodeum. The surrounding swellings are called the anal folds (Pansky, 1982).

2.1 Anatomy

The anal canal is about 4 cm in length and starts as continuation of the rectum, where it is angulated at nearly a right angle by the puborectalis muscle, causing an angulation of the anal canal posteriorly and downwards.

The anal canal is divided into two parts by the site of junction of the ectoderm and endoderm, an upper and lower part; the upper part derives its characteristics from its endodermal origin (Agur et al., 1999).

2.2 Upper Part of the Anal Canal

2.2.1 Histology

The lining is columnar epithelia thrown into nearly a dozen vertical columns (columns of Morgagni). These columns are connected at their lower ends by small horizontal folds, the anal valves (valves of Ball), which, together, form the pectinate line (Fig. 2). Above the valves, pockets are formed called anal

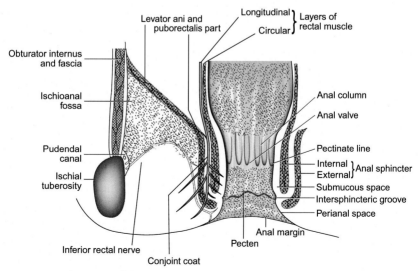

Fig. 2 Coronal section of the anal canal.

sinuses. Ten mucous-secreting anal glands open into these sinuses; they are submucosal glands.

2.2.2 Blood Supply

The anal canal derives its blood supply from the superior rectal vessels, the branches of which terminate within the anal columns, and ultimately, the inferior mesenteric vessels. A small part of the muscular wall is supplied by the middle rectal and median sacral arteries. The veins correspond to the above arteries, forming an internal rectal venous plexus which forms three cushion-like masses in the 3, 7, and 11 o'clock positions, corresponding to the common sites for hemorrhoids.

2.2.3 Lymph Vessels

Lymphatic drainage is to the pararectal nodes and eventually, the inferior mesenteric preaortic lymph nodes.

2.2.4 Nerve Supply

It is to the autonomic plexuses which are not sensitive to somatic painful stimuli. Sympathetic fibers from the pelvic plexus, with preganglionic cell bodies in L1 and 2 segments of the cord, cause contraction of the internal sphincter, and pelvic splanchnic (parasympathetic) nerves relax it. Afferent fibers from the upper end of the canal run mainly with the pelvic splanchnics (Sinnatamby, 1999).

2.3 The Lower Half of the Anal Canal

It starts below the pectinate line (sometimes called the dentate line) which is a smooth-surfaced area, and the pecten extends down to the intersphincteric groove.

2.3.1 Histology

There is no abrupt line of change from the columnar gut type to squamous pecten type. It is nonkeratinizing squamous epithelia devoid of hair follicles, sebaceous glands, and sweat glands. Below the intersphincteric groove is a truly cutaneous area (keratinizing with hair follicles, sweat, and sebaceous glands) and this is continuous at the anus (anal margin) with the skin of the buttocks.

2.3.2 Blood Supply

It is from local vessels like the internal pudendal artery and ultimately to the internal iliac vessels. The veins correspond to the above arteries.

2.3.3 Lymph Vessels

Lymphatic drainage is to the inguinal lymph nodes.

2.3.4 Nerve Supply

It is the pudendal nerve (S2) which is sensitive to somatic pain (Romanes, 1986).

2.4 Anal Sphincters

The sphincter mechanism of the anal canal is a rather complicated muscle arrangement consisting of an internal anal sphincter and an external anal sphincter. These sphincters, assisted by the configuration of the mucous membrane, make the anal canal closed at all times except for passage of flatus and feces. There are two sphincters as demonstrated in Fig. 3 (Fritsch et al., 2002).

1. **The internal anal sphincter** is involuntary muscle and is a continuation of the circular muscle of the rectum. It is responsible for 85% of the resting tone of the anal canal but contributes little to continence. That's why it can be divided without causing loss of continence (Papaconstantinou, 2005).

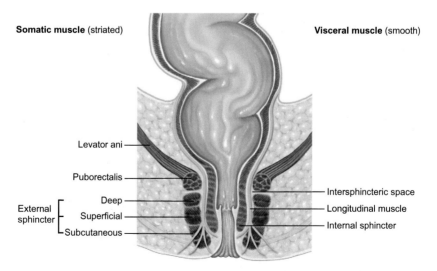

Somatic muscle (striated)   Visceral muscle (smooth)

Levator ani

Puborectalis

Deep   Intersphincteric space
External   Longitudinal muscle
sphincter   Superficial   Internal sphincter
Subcutaneous

Fig. 3 Types of anal sphincters. *(Windsor, A.C.J., Cohen, C.R.G., Campbell-Smith, T., 2012. Anal and related disorders. In: Henry, M.M., Thompson, J.M. (Eds.), Clinical Surgery, 3rd ed. Saunders-Elsevier, pp. 386 (Chapter 26).)*

2. **The external anal sphincter** is a voluntary muscle surrounding the internal sphincter and covering its lower edge, curving medially to reach close to the skin of the anal orifice. Its strong voluntary contractions prevent defecation and contribute to the major activity of continence. So, excessive division of the external sphincter can lead to fecal incontinence. It contributes only to 15% of the resting anal tone (Papaconstantinou, 2005). It consists of three parts:

   (a) **Subcutaneous part**: It is a ring of fibers that curves medially and lies below the lower end of the internal sphincter. This submucosal apposition of the two sphincters in the lower part of the canal gives rise to the palpable intersphincteric groove.

   (b) **Superficial part**: It is elliptical rather than circular, and it attaches to the tip of the coccyx at the back and to the perineal body in front, giving it bony attachment.

   (c) **Deep part**: The deep part blends with the internal sphincter and the levator ani muscle (except in the midline at the front where there are no levator ani fibers; here, the sphincter fibers alone complete the ring) to form the anorectal ring, which can be felt nearly 1 in. from the anal verge on rectal examination.

The longitudinal smooth muscles of the anal canal continue downward between the internal and external sphincter. Some of the fibers attach to the mucus membrane of the anal canal, while others pass laterally into the ishcioanal fossa, and the lower fibers traverse the external sphincter in a fan-shaped fashion, ending in the subcutaneous tissue of perianal skin (Gray and Lewis, 2000).

2.5 Relations

1. Posteriorly, anococcygeal body (fibrous tissue separating the anal canal) from the coccyx, laterally, the ischioanal fossa (which is more correct than what it was previously called, ishciorectal fossa) containing fat and the pudendal nerve and vessels.

2. Anteriorly, the perineal body separates it from the bulb of the urethra in the male and lower vagina in the female (Church et al., 1987).

3 THE PELVIC DIAPHRAGM

The lower pelvic aperture is closed by the pelvic diaphragm which consists of muscles and fascia (Fig. 4).

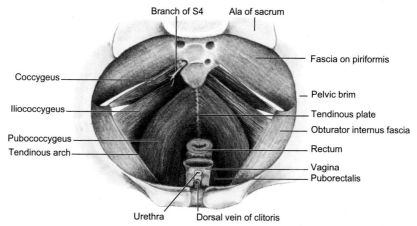

Fig. 4 Pelvic diaphragm. *(Chummy, S., Sinnatamby, 2011. Last's Anatomy: Regional and Applied. Churchill Livingstone, p. 290 (Chapter 5).)*

All the viscera must traverse this diaphragm to gain access to the exterior. These muscles consist of:

1. Levator ani and coccygeus (both form the pelvic diaphragm). The levator ani arises from posterior aspect of the pubic bones, fascial covering of obturator internus, and spine of ischium, so that it forms a series of loops.

   (a) Levator prostatae or vaginal sphincter (a sling around the prostate or vagina).

   (b) Puborectalis slings around the rectoanal junction, blending with the internal and external sphincters, forming the anorectal ring.

   (c) What remains is the most posterior fibers attached to the sides of the coccyx and to a median fibrous raphe, which stretches between the apex of the coccyx and the anorectal junction.

2. Superficial muscles of the anterior urogenital perineum and posterior anal perineum (Kaiser and Ortega, 2002).

REFERENCES

Agur, A.M.R., Lee, M.J., Grant, J.C.B., 1999. Grant's Atlas of Anatomy, tenth ed. Lippincott Williams and Wilkins, London.

Church, J.M., Raudkivi, P.J., Hill, G.L., 1987. The surgical anatomy of the rectum—a review with particular relevance to the hazards of rectal mobilisation. Int. J. Color. Dis. 2, 158–166.

Dobson, H.D., Pearl, R.K., Orsay, C.P., et al., 2003. Virtual reality: New method of teaching anorectal and pelvic floor anatomy. Dis. Colon Rectum 46, 349–352.

Fritsch, H., Brenner, E., Lienemann, A., Ludwikowski, B., 2002. Anal sphincter complex: reinterpreted morphology and its clinical relevance. Dis. Colon Rectum 45, 188–194.

Gray, H., Lewis, W.H., 2000. Gray's Anatomy of the Human Body, twentieth ed. Bartleby, New York, NY.

Kaiser, A.M., Ortega, A.E., 2002. Anorectal anatomy. Surg. Clin. North Am. 82, 1125–1138.

Pansky, B., 1982. Review of Medical Embryology. LifeMap Sciences Inc. (Chapter 7, pp. 84).

Papaconstantinou, H.T., 2005. Evaluation of anal incontinence: minimal approach, maximal effectiveness. Clin. Colon Rectal Surg. 18, 9–16.

Romanes, G.J., 1986. Thorax and Abdomen. Cunningham's Manual of Practical Anatomy, fifteenth ed. Medical Publications, Oxford University Press, New York, NY. vol. II.

Sinnatamby, C.S., 1999. Last's Anatomy: Regional and Applied, tenth ed. Churchill Livingstone, Edinburgh.

CHAPTER 2

Epidemiology of Perianal Disease

Batool Mutar Mahdi

Contents

1 INTRODUCTION

The actual incidence of perianal disease is underestimated, if we consider those perianal abscesses that end with spontaneous drainage or those treated in the emergency room, or even in the physician's clinic (Abcarian, 2011). Demographic studies point to a clear disparity in the occurrence of anal disease with respect to age, sex, and season, but no obvious pattern exists among various countries or regions of the world. Although it has been suggested that there is a direct relation between the formation of anorectal diseases and bowel habits, frequent diarrhea, and poor personal hygiene, this relation remains unproven.

2 SEASON-RELATED DEMOGRAPHICS

The occurrence and predisposition of perianal disease differs with the seasons of the year. The incidence of abscess formation appears to be higher in spring and summer (Abcarian, 2011). Other countries, however, reported a higher prevalence in June and a lower incidence in August and September (Vasilevsky and Gordon, 1984). In Brazil, there was a higher incidence (61.5% of diagnoses) of perianal abscess in the summer and spring months (Netoa et al., 2016).

New Concepts in the Management of Septic Perianal Conditions
https://doi.org/10.1016/B978-0-12-816111-1.00002-1

3 COUNTRY-RELATED DEMOGRAPHICS

Perianal suppurations have an incidence of 1–2:10,000 inhabitants per year and represent about 5% of all proctology consultations (Martins and Pereira, 2010). In the United States, the estimated incidence is between 68,000 and 96,000 cases per year (Akkapulu et al., 2015). The reported incidence of anal fistula in patients with an anorectal abscess varies from 7.6% to 66.2% (Henrichsen and Christiansen, 1986) while others reported that in approximately 30%–50% of patients with an anorectal abscess, a persistent fistula in ano will develop (Hämäläinen and Sainio, 1998). Patients with perianal abscesses who had undergone surgery still had a propensity to develop anal fistula with an incidence rate of 11% (Wolff et al., 2007). During the follow-up period, the incidence of fistula development following management of an acute perianal abscess was 9.7%, while in other countries, it ranged from 5.77% to 85% (Nicholls et al., 1990). Other studies showed no recurrence (Inceoglu and Gencosmanoglu, 2003).

4 AGE- AND SEX-RELATED DEMOGRAPHICS

The peak incidence of anorectal abscesses is in the third and fourth decades of life. These abscesses are also quite common in infants and children (Novotny et al., 2008). The exact mechanism is poorly understood but does not appear to be related to constipation. Fortunately, this condition is quite benign in infants, rarely necessitating any operative intervention other than simple drainage. Men are affected more frequently than women, with a male-to-female predominance of 2:1 to 3:1 (Beard and Osborn, 2011). The age of most patients with anorectal suppuration is between 20 and 60 years, with a mean of 40 years and an incidence twice as high in men, reaching up to 83.9% of cases (Sözener et al., 2011). In Brazil, the mean age was 43.03 years and the condition occurred more frequently in males (73.1%) than in females (Netoa et al., 2016). It has been found that gender and age did not affect the recurrence after the first surgery (Akkapulu et al., 2015). Regarding incidence of chronic anal fistula or recurrent sepsis, it was 36.5% in patients whose ages were <40 years (Hamadani et al., 2009). The incidence of perianal fistula in male and female was 9.4% and 11.1%, respectively, in Sudan (Idris et al., 2011).

REFERENCES

Abcarian, H., 2011. Anorectal infection: abscess-fistula. Clin. Colon Rectal Surg. 24, 14–21.

Akkapulu, N., Dere, Ö., Zaim, G., Soy, H.E.A., Özmen, T., Dogrul, A.B., 2015. A retrospective analysis of 93 cases with anorectal abscess in a rural state hospital. Ulus. Cerrahi Derg. 31, 5–8.

Beard, J.M., Osborn, J., 2011. Anorectal abscess. In: Rakel, R.E., Rakel, D.P. (Eds.), Textbook of Family Medicine, eighth ed. Saunders, Philadelphia, PA.

Hamadani, A., Haigh, P.I., Liu, I.L., Abbas, M.A., 2009. Who is at risk for developing chronic anal fistula or recurrent anal sepsis after initial perianal abscess? Dis. Colon Rectum 52, 217–221.

Hämäläinen, K.P., Sainio, A.P., 1998. Incidence of fistulas after drainage of acute anorectal abscesses. Dis. Colon Rectum 41, 1357–1361.

Henrichsen, S., Christiansen, J., 1986. Incidence of fistula-in-ano complicating anorectal sepsis: a prospective study. Br. J. Surg. 73, 371–372.

Idris, S.A., Hamza, A.A., Alegail, I.M.A., 2011. The relation between the presence of intestinal bacteria in the perianal abscess and the anticipated perianal fistula. Sudan JMS 6, 199–208.

Inceoglu, R., Gencosmanoglu, R., 2003. Fistulotomy and drainage of deep postanal space abscess in the treatment of posterior horseshoe fistula. BMC Surg. 3, 10.

Martins, I., Pereira, J.C., 2010. Supurações perianais. Rev. Port. Coloproctol. 7, 118–124.

Netoa, I.J.F.C., Werckaa, J., Cecchinnia, A.R.S., Lopesa, E.A., Wattéa, H.H., et al., 2016. Perianal abscess: a descriptive analysis of cases treated at the Hospital Santa Marcelina, São Paulo. J. Coloproctol. (Rio J.) 36, 149–152.

Nicholls, G., Heaton, N.D., Lewis, A.M., 1990. Use of bacteriology in anorectal sepsis as an indicator of anal fistula: experience in a distinct general hospital. J. R. Soc. Med. 83, 625–626.

Novotny, N.M., Mann, M.J., Rescorla, F.J., 2008. Fistula in ano in infants: who recurs? Pediatr. Surg. Int. 24, 1197–1199.

Sözener, U., Gedik, E., Kessaf Aslar, A., Ergun, H., Halil Elhan, A., Memikoğlu, O., et al., 2011. Does adjuvant antibiotic treatment after drainage of anorectal abscess prevent development of anal fistulas? A randomized, placebo-controlled, double-blind, multicenter study. Dis. Colon Rectum 54, 923–929.

Vasilevsky, C.A., Gordon, P.H., 1984. The incidence of recurrent abscesses or fistula-in-ano following anorectal suppuration. Dis. Colon Rectum 27, 126–130.

Wolff, B.G., Fleshman, J.W., Beck, D.E., Pemberton, J.H., Wexner, S.D. (Eds.), 2007. The ASCRS Text Book of Colon and Rectal Surgery. Springer, New York, pp. 195–196.

CHAPTER 3

Pathogenesis of Anal Diseases

Riyadh Mohammad Hasan

Contents

1 INTRODUCTION

The anorectum is surrounded by several potential spaces filled with areolar tissue. As shown in Fig. 1, they are:

1. The perianal space surrounding the anus.
2. The ischioanal space lying between the anal canal and ischium bound superiorly by the levator ani.
3. Both ischioanal spaces which are connected posteriorly by the deep post-anal space or Courtney's space, bound by the levator ani superiorly and anococcygeal ligament below.
4. The intersphincteric space lying between the internal and external sphincters which continues below with the perianal space and above with the wall of the rectum.
5. The supralevator space above the levator ani, which communicates posteriorly allowing spread up to the retroperitoneum (Klosterhalfen et al., 1991).

Perianal abscesses and fistulas that occur in the anorectum area represent about 90% of anorectal disorders. The pathogenesis of abscesses and fistulas is usually the same; the acute phase is represented by perianal abscess, and the chronic phase by the fistula, arising mainly from the obstruction of anal crypts (Rizzo et al., 2010) associated with increased sphincter tone (Hamadani et al., 2009). Infection of the static glandular secretions by aerobic

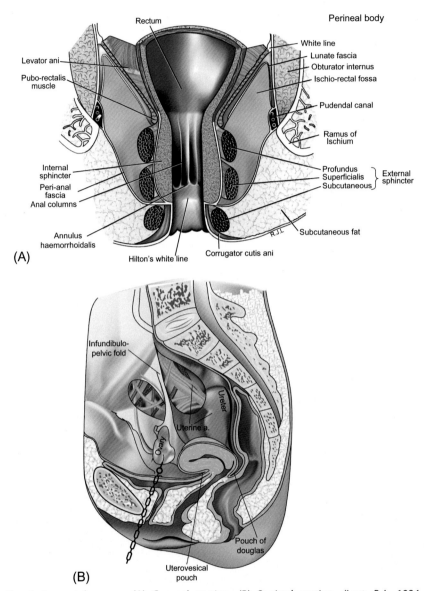

Fig. 1 Anorectal spaces. (A) Coronal section. (B) Sagittal section. *(Last, R.J., 1984. Anatomy Regional and Applied, seventh ed. Churchil, Livingstone, pp. 337, 349 (Section 5)).*

and anaerobic bacteria results in suppuration and abscess formation within the anal gland (Brook and Frazier, 1997). Typically, the abscess initially forms in the intersphincteric space and then spreads along adjacent potential spaces (Parks, 1961).

2 PATHOGENESIS OF ANAL ABSCESS

There is an occurrence of a simple boil and skin appendage infection occurring at the anal region. Perianal abscess and fistula are walled-off infections occurring most frequently due to infection of the anal glands, about a dozen glands opening in the anal crypts at the dentate line (which is called the cryptoglandular etiology), accounting for about 90% of cases (Klosterhalfen et al., 1991). The anal canal has 6–14 glands that lie in the plane between the internal and external anal sphincters (Abeysuriya et al., 2010). Ducts from these glands pass through the internal sphincters and drain into the anal crypts at the dentate line (Abeysuriya et al., 2010). If the crypt does not spontaneously drain into the anal canal, an infection of the intersphincteric space may occur. This infection may spread along the intersphincteric space and result in an intersphincteric, perianal, or supralevator abscess (Fig. 2A). The infection may also pass through the external anal sphincter and result in a perirectal abscess (Ramanujam et al., 1984; Sahnan et al., 2017) (Fig. 2B).

These glands lie in the intersphincteric space, so the intersphincteric abscess forms in this space. Then, it may extend upward, downward, or circumferentially, forming the ischioanal or supralevator abscesses (Whiteford, 2007).

The other 10% of causes are:

1. Inflammatory bowel disease, especially Crohn's disease, and this diagnosis must be considered in patients with recurrent anorectal abscesses (Tio et al., 1990; Fish and Kugathasan, 2004; Zhu et al., 2016).
2. Infection (tuberculosis, actinomycosis, lymphogranuloma venereum) (Coremans et al., 2005).
3. Trauma and radiation (Bennetsen, 2008).
4. Postoperative: episiotomy, hemorrhoidectomy, prostatectomy (Vasilevsky and Gordon, 2007).
5. Malignancy, carcinoma, leukemia, lymphoma (Vanheuverzwyn et al., 1980; Barnes et al., 1984; Nelson et al., 1985)

These abscesses are named after the space they fill, but if we put them in the order of frequency, we have perianal abscess (the most common), then ischioanal, intersphincteric, and supralevator (the rarest). A horseshoe abscess is the spread of abscess from one ischioanal fossa to the other through the deep postanal space. A supralevator abscess may result from extension of an intersphincteric or ischioanal abscess upward or extension of pelvic abscess downward (Ommer et al., 2012).

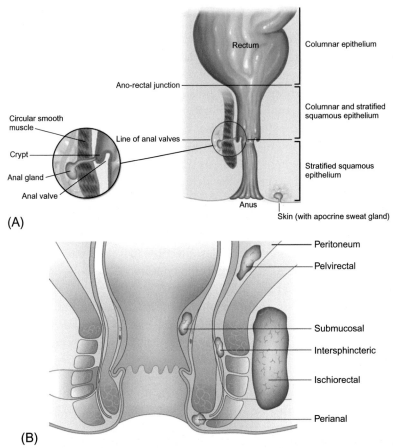

Fig. 2 (A) Common cause of anal abscess. (B) Classification of anal abscesses. (A) *(Windsor, A.C., 2010. Anal and related disorders. In: Henry, M.M., Thompson, J.M. (Eds.), Clinical Surgery, second ed. Saunders-Elesvier, pp. 414 (Chapter 25)). (B) (Fazio, V.W., Maher, J.W., Williamson, R.C.N., 2001. Gastrointestinal surgery. In: Corson, J.D., Williamson, R.C.N. (Eds.), Surgery, Mosby, pp. 21:10 (Section 3, Chapter 21)).*

Anorectal abscesses are associated with anal fistulas in 37% of patients. If these fistulas are not recognized and treated, perirectal abscesses may recur (Vasilevsky and Gordon, 1984).

3 PATHOGENESIS OF ANAL FISTULA

Perianal fistula is a communication between the anal canal and the perianal skin and is most commonly the sequel of drainage of perianal abscess (Fig. 3).

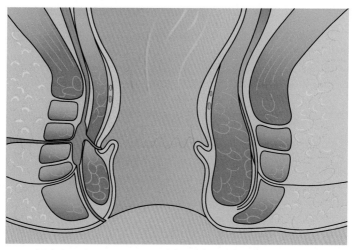

Fig. 3 Perianal fistula is a communication between the anal canal and the perianal skin and is the sequel of drainage of perianal abscess. *(Fazio, V.W., Maher, J.W., Williamson, R.C.N., 2001. Gastrointestinal surgery. In: Corson, J.D., Williamson, R.C.N. (Eds.), Surgery, Mosby, pp. 21:11 (Section 3, Chapter 21)).*

Most anal fistulas originate in anal crypts, which become infected, with ensuing abscess formation. When the abscess is opened or it ruptures, a fistula is formed. An anal fistula can have multiple accessory tracts complicating its anatomy (Wald et al., 2014).

The chronic fistula-in-ano represents a pathogenic factor in the genesis of anal fistula, especially in the posterior part of the anal canal. In this region, we found trans- and suprasphincteric tracks with abscesses in the fossa ischioanal (Jostarndt et al., 1984).

Other causes of anal fistulas include opened perianal or ischioanal abscesses, which drain spontaneously through these fistulous tracts. Fistulas are also found in patients with inflammatory bowel disease, particularly Crohn's disease (Garcia-Aguilar et al., 1996). Perianal activity often parallels abdominal disease activity, but it may occasionally be the primary site of active disease.

Anal fistulas can also be associated with diverticulitis, foreign-body reactions, actinomycosis, chlamydia, lymphogranuloma venereum (LGV), syphilis, tuberculosis (Chung et al., 1997), radiation exposure, and HIV disease. Approximately 30% of patients with HIV disease develop anorectal abscesses and fistulas (North Jr et al., 1996).

Fistulas are also classified according to their relation to the anal sphincters. Anal fistulas are classified into the following four general types (Abbas et al., 2011) (Fig. 4)

1. Intersphincteric—Through the dentate line to the anal verge, tracking along the intersphincteric plane, ending in the perianal skin.

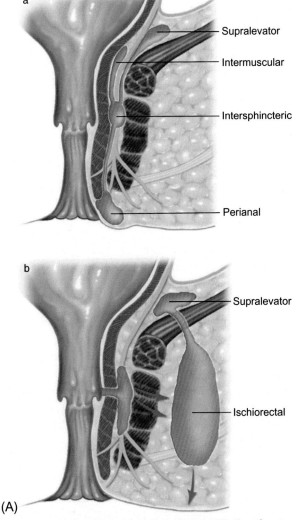

Fig. 4 (A) Vertical and horizontal spread of the infection. (B) Classification of fistula-in-ano. (a) Intersphincteric. (b) Transsphincteric. (c) Suprasphincteric/Extrasphincteric. (A) *(Windsor, A.C., 2010. Anal and related disorders. In: Henry, M.M., Thompson, J.M., (Eds.), Clinical Surgery, second ed. Saunders-Elesvier, pp. 424 (Chapter 25)). (B) (Windsor, A.C., (2010). Anal and related disorders. In: Henry, M.M., Thompson, J.M., (Eds.), Clinical Surgery, second ed. Saunders-Elesvier, pp. 425 (Chapter 25)).*

(Continued)

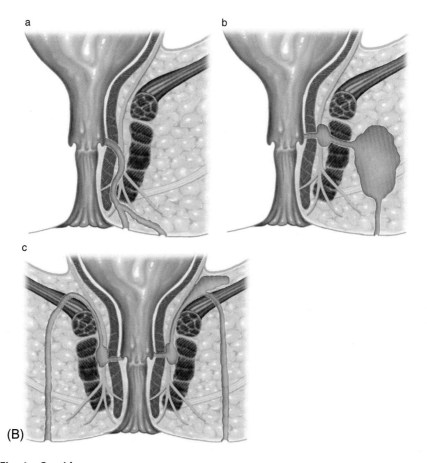

(B)

Fig. 4—Cont'd

2. Transsphincteric—Through the external sphincter into the ischioanal fossa, encompassing a portion of the internal and external sphincter, ending in the skin overlying buttocks.
3. Suprasphincteric—Through the anal crypt and encircling the entire sphincter, ending in the ischioanal fossa.
4. Extrasphincteric—Starting high in the anal canal, encompassing the entire sphincter and ending in the skin overlying the buttocks.

Thus, fistula-in-ano is the most common form of perineal abscess and they are comprised of an internal opening, an external opening, and a track between them. The external opening appears following infection and/or an abscess.

REFERENCES

Abbas, M.A., Jackson, C.H., Haigh, P.I., 2011. Predictors of outcome for anal fistula surgery. Arch. Surg. 146, 1011–1016.

Abeysuriya, V., Salgado, L.S., Samarasekera, D.N., 2010. The distribution of the anal glands and the variable regional occurrence of fistula-in-ano: is there a relationship? Tech. Coloproctol. 14, 317–321.

Barnes, S.G., Sattler, F.R., Ballard, J.O., 1984. Perirectal infections in acute leukemia. Improved survival after incision and debridement. Ann. Intern. Med. 100, 515–516.

Beck, D.E., 2003. Hand Book of Colorectal Surgery, second ed. Marcel Dekker Inc.

Bennetsen, D.T., 2008. Perirectal abscess after accidental toothpick ingestion. J. Emerg. Med. 34, 203–204.

Brook, I., Frazier, E.H., 1997. The aerobic and anaerobic bacteriology of perirectal abscesses. J. Clin. Microbiol. 35, 2974–2976.

Chung, C.C., Choi, C.L., Kwok, S.P., Leung, K.L., Lau, W.Y., Li, A.K., 1997. Anal and perianal tuberculosis: a report of three cases in 10 years. J. R. Coll. Surg. Edinb. 42, 189–190.

Coremans, G., Margaritis, V., Van Poppel, H.P., Christiaens, M., Gruwez, J., Geboes, K., Wyndaele, J., Vanbeckevoort, D., Janssens, J., 2005. Actinomycosis, a rare and unsuspected cause of anal fistulous abscess: report of three cases and review of the literature. Dis. Colon Rectum 48, 575–581.

Fazio, V.W., Maher, J.W., Williamson, R.C.N., 2001. Gastrointestinal surgery. In: Corson, J.D., Williamson, R.C.N. (Eds.), Surgery. Mosby, pp. 21:10, 21:11 (Section 3, Chapter 21).

Fish, D., Kugathasan, S., 2004. Inflammatory bowel disease. Adolesc. Med Clin. 15, 67–90.

Garcia-Aguilar, J., Belmonte, C., Wong, W.D., Goldberg, S.M., Madoff, R.D., 1996. Anal fistula surgery. Factors associated with recurrence and incontinence. Dis. Colon Rectum 39, 723–729.

Hamadani, A., Haigh, P.I., Liu, I.L., Abbas, M.A., 2009. Who is at risk for developing chronic anal fistula or recurrent anal sepsis after initial perianal abscess? Dis. Colon Rectum 52, 217–221.

Jostarndt, L., Nitsche, D., Thiede, A., Schröder, D., 1984. Pathogenesis and morphology of anal fistulas. Fortschr. Med. 14, 615–618.

Klosterhalfen, B., Offner, F., Vogel, P., Kirkpatrick, C.J., 1991. Anatomic nature and surgical significance of anal sinus and anal intramuscular glands. Dis. Colon Rectum 34, 156–160.

Last, R.J., 1984. Anatomy Regional and Applied, seventh ed. Churchil, Livingstone, pp. 337, 349 (Section 5).

Nelson, R.L., Prasad, M.L., Abcarian, H., 1985. Anal carcinoma presenting as a perirectal abscess or fistula. Arch. Surg. 120, 632–635.

North Jr., J.H., Weber, T.K., Rodriguez-Bigas, M.A., Meropol, N.J., Petrelli, N.J., 1996. The management of infectious and noninfectious anorectal complications in patients with leukemia. J. Am. Coll. Surg. 183, 322–328.

Ommer, A., Herold, A., Berg, E., Fürst, A., Sailer, M., Schiedeck, T., 2012. German S3 guideline: anal abscess. Int. J. Color. Dis. 27, 831–837.

Parks, A.G., 1961. Pathogenesis and treatment of fistula-in-ano. Br. Med. J. 1, 463–469.

Ramanujam, P.S., Prasad, M.L., Abcarian, H., Tan, A.B., 1984. Perianal abscesses and fistulas. A study of 1023 patients. Dis. Colon Rectum 27, 593–597.

Rizzo, J.A., Naig, A.L., Johnson, E.K., 2010. Anorectal abscess and fistula-in-ano: evidence-based management. Surg. Clin. North Am. 90, 45–68.

Sahnan, K., Adegbola, S.O., Tozer, P.J., Watfah, J., Phillips, R.K., 2017. Perianal abscess. BMJ 356. j475.

Tio, T.L., Mulder, C.J., Wijers, O.B., et al., 1990. Endosonography of peri-anal and peri-colorectal fistula and/or abscess in Crohn's disease. Gastrointest. Endosc. 36, 331–336.

Vanheuverzwyn, R., Delannoy, A., Michaux, J.L., Dive, C., 1980. Anal lesions in hematologic diseases. Dis. Colon Rectum 23, 310–312.

Vasilevsky, C.A., 2003. Anorectal abscess and fistula-in-ano. In: Beck, D. (Ed.), Handbook of Colorectal Surgery, second ed. Taylor & Francis Group LLC.

Vasilevsky, C.A., Gordon, P.H., 1984. The incidence of recurrent abscesses or fistula-in-ano following anorectal suppuration. Dis. Colon Rectum 27, 126–130.

Vasilevsky, C.A., Gordon, P.H., 2007. Benign anorectal: abscess and fistula. In: The ASCRS Textbook of Colon and Rectal Surgery. Springer, New York, pp. 192–214.

Wald, A., Bharucha, A.E., Cosman, B.C., Whitehead, W.E., 2014. ACG clinical guideline: management of benign anorectal disorders. Am. J. Gastroenterol. 109, 1141–1157.

Whiteford, M.H., 2007. Perianal abscess/fistula disease. Clin. Colon Rectal Surg. 20, 102–109.

Windsor, A.C., 2010. Anal and related disorders. In: Henry, M.M., Thompson, J.M. (Eds.), Clinical Surgery, second ed. Saunders-Elsevier, pp. 414, 424, 425 (Chapter 25).

Zhu, P., Chen, Y., Gu, Y., Chen, H., An, X., Cheng, Y., Gao, Y., Yang, B., 2016. Analysis of clinical characteristics of perianal Crohn's disease in a single-center. Zhonghua Wei Chang Wai Ke Za Zhi 19, 1384–1388.

FURTHER READING

Nordgren, S., Fasth, S., Hulten, L., 1992. Anal fistulas in Crohn's disease: incidence and outcome of surgical treatment. Int. J. Color. Dis. 7, 214–218.

CHAPTER 4

Immune Response to Perianal Abscess

Batool Mutar Mahdi

Contents

1 INTRODUCTION

Perianal abscess (PA) is defined as a cavity filled with pus around either the anus or anus and rectum. The main patient complaints are pain around the anus, swelling, and redness around the anus associated with fever and chills (Whiteford et al., 2005). This abscess either ruptures spontaneously or is evacuated by surgical operation (incision and drainage) under either local or general anesthesia, providing complete relief (Malik et al., 2010). The most common cause of this devastating abscess is anal gland infection, in which one or more of the seven to eight small glands inside the anus becomes impacted with fecal material, in a process called cryptoglandular hypothesis (Parks, 1961). Other rare causes include sexually transmitted diseases, hair follicle infections, sweat gland infections, inflammatory bowel disease, and compromised immune system (Grace, 1990). Thus, the immune system is a powerful defense mechanism against development of this disease.

2 IMMUNE RESPONSE TO MICROBIAL INFECTION

The immune response is divided into two lines of defense mechanisms, innate and adaptive immunity. Innate immunity is the first line of defense and includes anatomical barriers (represented by skin and mucous

New Concepts in the Management of Septic Perianal Conditions
https://doi.org/10.1016/B978-0-12-816111-1.00004-5

membrane) preventing pathogen entry, chemical barriers (like hydrochloric acid in the stomach and enzymes in tears and saliva), and physiological barriers (represented by high temperature during infection) (Corrales et al., 2016). The last component of innate immunity is the cellular barrier, which includes phagocytic cells like macrophages, which the body's scavengers which, in cooperation with neutrophils and natural killer cells, remove any foreign antigenic materials that enter the body. Macrophages are typically divided into the classically activated M1 and activated M2 subsets, which are referred to as inflammatory and wound-healing macrophages, respectively. The recognition of these microbial antigens is achieved by interaction of the toll-like receptor (TLR) with pathogen-associated molecular pattern (PAMP), and a macrophage with M1 phenotype occurs in response to TLR ligands (LPS), leading to activation of multiple signal transductions inside immune cells, resulting in the production of cytokines and chemokines. These mediators recruit immune cells to the site of infection. These immune cells kill pathogens through the induction of reactive oxygen radicals, reactive nitrogen (like nitric oxide), and antimicrobial peptides and proteins (Scholtzova et al., 2016). Besides the fragmentation of antigens, this antigenic peptide requires further processing through presentation of the molecules to CD4 T cells by MHC Class II molecules, inducing CD4+ T cell proliferation and differentiation (Duan et al., 2008). MHC proteins such as HLA-DR, HLA-DQ, HLA-DM, and the invariant chain (Ii) are assembled in the endoplasmic reticulum (ER) and are transported through the Golgi apparatus, before entering exocytic vesicles. The antigenic peptide (endosome/lysosomes) then fuse with the exocytic vesicles, leading to the formation of the MHC II vesicles containing HLA-DR, HLA-DQ, HLA-DM, LAMP-1, and antigenic peptide. During this process, acidification of MHC II activates proteases that cleave protein antigens including Ii. This process leads to an exchange of antigens with Class II-associated invariant chain peptides (CLIP). Processed, fragmented antigens are then loaded onto MHC II molecules, followed by transportation to the surface of antigen-presenting cells that activate CD4+ T cells (Mazmanian et al., 2005).

If the first line of defense fails, then the second line of protection (adaptive immunity) will be activated in response to specific antigens, because this line of defense is characterized by specificity, memory, and clonal expansion. Its activation occurs through a number of complex signals between antigen-presenting cells (macrophage and dendritic cells) and T lymphocytes (Hilchie et al., 2013). There are three types of T cells (T Helper [T_H],

T cytotoxic [T_c], and T regulatory cells [T_{reg}]) (Corthay, 2009). After activation, T_H cells secrete cytokines that lead to the evolution of the T_H9, T_H17, T_H22, $T_{reg}1$, and induced T_{reg} (iT_{reg}) cell responses, but traditionally, T cells have usually been divided into either a T_H1 (cell mediated immunity) that secretes IFN-γ, TNF-α, and GM-CSF or a T_H2 cell response (humoral immunity) by B-cell activation and antibodies production (Annunziato and Romagnani, 2009). T_c cells will bind to and kill infected cells, such as those that are infected with viruses, or mutated host cells like cancer cells. T_{reg} cells generally modulate these processes. After maturation, macrophages are activated in a dynamic response to pathogens or other stimuli to acquire specialized functional phenotypes. As for the lymphocyte system, a dichotomy has been proposed for macrophage activation: classic and alternative, or M1 and M2, respectively. M1 macrophages produce proinflammatory cytokines like TNF-α and IL-13 and enhanced microbicidal and antigen-presentation activities. In contrast, M2 phenotype macrophages respond to T_H2 cytokines such as IL-4 and IL-10 and M-CSF. M2 macrophages enhance the expression of antiinflammatory cytokines (IL-10), which suppress the expression of proinflammatory cytokines, thereby maintaining immunological homeostasis, and enhanced phagocytic activity (Martinez and Gordon, 2014). Histopathological studies of anal fistulas showed epithelialized tissues surrounded by dense collagen tissue with compact inflammatory cells. Anal fistulas usually develop in the presence of a perianal abscess that contains gut flora. These imply a microbiological cause, but bacteria are infrequently found in chronic fistulas. The increased expression of proinflammatory cytokines and epithelial to mesenchymal cell transition are demonstrated in both cryptoglandular and Crohn's perianal fistulas. This suggests that molecular mechanisms may also play a role in both fistula development and persistence (Sugrue et al., 2017).

3 IMMUNE RESPONSE TO PERIANAL ABSCESS IN IMMUNE COMPETENCE PATIENTS

Patients with normal immune response have normal T and B-cells in number, the neutrophil function evaluated with oxidative burst test was also normal, and in vitro T cell proliferation response to mitogens (PHA) is normal. When infection is initiated, it leads to stimulation of inflammatory processes. The first line of defense is macrophages, showing that upregulated costimulatory molecules like CD80/CD86 and secretion of cytokine profile leads to Type I proinflammatory response. The release of interleukin

2 (IL-2) and interferon-γ (IFN-γ) by naive T lymphocytes predominantly stimulates cytotoxic T lymphocytes, macrophages, and natural killer (NK) cells and increases the antigen-presenting potential of all these cell types. Contradictory to this proinflammatory immune reaction, a compensatory antiinflammatory response has occurred involving the secretion of interleukin 4 (IL-4) and interleukin 10 (IL-10) that down-regulate inflammatory mediators including tumor necrosis factor-α (TNF-α) and interleukin 1 (IL-1), and favor a humoral immune response. The combined effects of IL-4 and IL-10 have been shown to shift the Th1/Th2 cell activation in favor of a Th2 immune response, which seems to be essential for fighting against the infection, inflammation, and final healing (Rogy et al., 2004). Interleukin-10 plays an important role as an antiinflammatory cytokine in humans and its deficiency or its receptor (IL-10/IL-10R) leads to defective STAT3 dimerization and causes severe dysregulation of the immune system resulting in perianal disease (Loddo and Romano, 2015). IL-10 is an antiinflammatory cytokine produced by many cells like T_{reg}, macrophages, granulocytes, and dendritic cells. IL-10 has a critical role in maintaining the balance of the immune system (immune homeostasis) by inhibiting the secretion of T helper 1 cytokines (IL-2 and IFN-γ). IL-10 provides protection against excessive immune responses and tissue damage caused by infection and inflammation (Glocker et al., 2011).

In addition, susceptibility to infection can be affected by microbiota and the bacterial community in the gut. Polysaccharides (PS) of *Bacteroides fragilis*, which is an intestinal commensal bacterium, can initiate both pro-inflammatory and immunomodulatory responses (Chang et al., 2016). PS may change the equilibrium between DCs, Th17 cells, and IL-10-producing regulatory T cells (Surana and Kasper, 2012). Under normobiosis, *B. fragilis* protects against colitis and other inflammation through production of IL-10 by CD4+ T helper cells. When intestinal integrity is disrupted or under dysbiosis, *B. fragilis* and intestinal bacteria may escape into other parts of the body, resulting in abscess formation. The molecular mechanism used by *B. fragilis* shares common pathways (PS-TLR2) (toll-like receptor-2) and interactions between *B. fragilis* (especially its PS fraction) and immune cells (dendritic cell-specific intercellular adhesion molecule-3-grabbing nonintegrin) receptor on human DCs, which is also important for PSA processing and presentation to T cells. This interaction plays an important role for the development of infection by inducting a complex signaling network that finally modulates innate and adaptive immunity in the host (Chang et al., 2016).

4 IMMUNE RESPONSE TO PERIANAL ABSCESS IN IMMUNE-COMPROMISED PATIENTS

Many immunodeficiency disorders either primary (like selective IgA deficiency, DiGeorge syndrome, common variable immunodeficiency, Bruton's disease, immune dysregulation polyendocrinopathy-enteropathy X-linked [IPEX] syndrome, Wiskott-Aldrich syndrome, chronic granulomatous disease, and severe combined immunodeficiency disorders [SCID]) or secondary immune deficiency (like diabetes mellitus, immune-mediated encompassing staphylococcal skin carriage, chronic skin disease, cancer, HIV infection) are associated with gastrointestinal symptoms like chronic diarrhea, nodular lymphoid hyperplasia, malabsorption, and perianal abscess; psychogenic causes are also associated with factitious abscess formation (Bousfiha et al., 2015). One immune genetic cause of recurrent infections and perianal lesions and abscess is IL-10/IL-10R deficiencies (Karaca et al., 2016). Thus, microbial infection and defects in the immune system are regarded as predisposing factors for abscess formation (Smink et al., 2011). The infection arises from inoculation of different bacteria in the perineal area, facilitated by an impairment in the immune system like diabetes mellitus, alcoholism, malignancies (including hematological malignancies like leukemia), treatment with corticosteroids, HIV infection, renal failure, and hemodialysis (Nada et al., 2007). Patients with diabetes mellitus had defective phagocytosis and were prone to increased incidence of urinary tract infections as a result of functional urinary tract obstructions from diabetic neuropathy and disease of the small vessels (Nisbet and Thompson, 2002). The first line of defense is neutrophils, the type of innate immunity that is recruited with IL-1β/IL-1R activation by TLR2, NOD2, FPR1, and the ASC/NLRP3 inflammasome in an α-toxin-dependent mechanism. Deficiency in IL-1β leads to neutrophil abscess formation in *Staph aureus* infection (Cho et al., 2012). In addition, neutropenic patients are unable to mount an adequate inflammatory response and are susceptible to infection by less virulent bacteria; consequently, they may have unusual presentations of infection, poor wound healing, and abscess formation (Bodey, 2000). Another important cell in the defense mechanism is the macrophage, which eliminates apoptotic cells and pathogens with the aid of cytokines that mediate the transition from innate to adaptive immunity and direct the fate of macrophages transitioning from inflammation-activated cells to antiinflammatory or "alternatively activated" macrophages. So, deregulated cytokine secretion is implicated in several disease states ranging from chronic inflammation to abscess formation (Arango Duque and Descoteaux, 2014). Thus, the majority of patients with an underlying primary immune deficiency had recurrent mucocutaneous abscesses that were associated with significant morbidity and long-term complications, including scarring and fistula formation. Failure to respond to antimicrobial therapy raises the possibility of immune deficiency. The infection site and the location of the abscess may give clues to the cause of the disease, like perianal or enterocutaneous abscess formation, which suggests underlying Crohn's disease rather than primary immune deficiency. Many people suffer from conditions associated with immune deficiency and acquire pathological abscess formation, like patients with high staphylococcal colonization rates in type 1 DM, intravenous drug users, patients on hemodialysis, patients with quantitative or qualitative defects in leukocyte function and in surgical patients. The type of bacteria is also important for example strains of *S. aureus* harboring the *lukS-lukF* gene encoding the Pantin-Valentine leucocidin (a leucocytotoxic toxin) are frequently associated with severe infection in persons with no known immune deficiency (Wiese-Posselt et al., 2007). The prevalence of infection and abscess formation in diabetic patients is due to decreased chemotaxis and phagocytosis of polymorphs, which is most marked when glycemic control is poor, or when there is vascular disease insufficiency, tissue hypoxia, and nerve damage (Pozzilli and Leslie, 1994). Microangiopathy may prevent adequate antibiotic penetration, leading to persistent soft tissue infection and necrotizing fasciitis, which is potentially one of the most serious effects. Inflammatory bowel disease like Crohn's disease is associated with enterocutaneous abscess and fistula involving the perineum and abdominal wall. The response to treatment was poor and required immune therapy with infliximab (Sands et al., 2004).

Other primary immune deficiency is Chronic granulomatous disease which is an inherited immune deficiency caused by defects in the nicotinamide adenine dinucleotide phosphate (NADPH) oxidase that generates superoxide in phagocytes (Wood et al., 2007). It is caused by mutations in any of the four structural genes of the NADPH oxidase, gp91phox, p22phox, p47phox or p67phox. The majority of cases are X-linked recessive, caused by mutations in gp91phox; the remainder are autosomal recessive. The child usually had recurrent infection or granulomatous disease in the skin, lymph nodes, lungs, liver and bones (Rosenzweig and Holland, 2004). In this disease, there is excessive production of IL-8 by neutrophils, delayed apoptosis. Both factors may contribute to the granulomatous inflammation (Malech and Hickstein, 2007). Also, there is an absence of active

NADPH oxidase in T lymphocytes leads to shift in T cells to a T helper 1 (Th-1) cytokine pathway on activation that increased risk of certain Th-1 autoimmune disorders in CGD patients.

Another disease affecting adhesion molecules of the leukocyte is leukocyte-adhesion deficiency syndrome-1 (LAD-1), an autosomal recessive disorder which results from mutations in the common chain of the b2 integrin family, CD18, affecting the b2 integrin heterodimers CD11a/CD18, CD11b/CD18, and CD11c/CD18. Neutrophils are unable to aggregate, and they bind to intercellular adhesion molecules on endothelial cells and migrate to sites of infection/inflammation, leading to infection and abscess formation (Rosenzweig and Holland, 2004). Another type of LAD is leucocyte-adhesion deficiency syndrome-2, which is a very rare autosomal recessive disorder caused by an inborn error of fucose metabolism, resulting in the absence of fucosylated glycans at the cell surface. The last type is LAD-3, which is characterized by the general failure to activate a number of integrins in response to cytokines or due to abnormality of the Rap-1 function, a regulatory GTPase involved in regulation of integrin signaling (Malech and Hickstein, 2007).

Another type of defect is mutation in WAS protein; the gene encodes a cytoplasmic protein which is crucial to cell polarization, migration, and phagocytosis (Thrasher et al., 2000). This results in Wiskott-Aldrich syndrome, which is characterized by recurrent infection and abscess formation.

An additional, related disease is severe congenital neutropenia (SCN), which is characterized by profound neutropenia resulting from failure of promyelocytes to mature into myelocytes. Infections include cellulitis, perirectal abscess, peritonitis, stomatitis, and meningitis. The main causes which have been identified to date is mutations in the genes encoding elastase 2 (ELA2), growth factor 11, and granulocyte-colony stimulating factor receptors (Notarangelo et al., 2006). Cold abscesses are other types of abscesses that formed in patients with the hyper-IgE syndrome due to abnormal neutrophil chemotaxis and an altered immunoglobulin profile as the primary immune defect. Missense mutations and single-codon in-frame deletions in *STAT3* have been reported in this disease (DeWitt et al., 2006).

REFERENCES

Annunziato, F., Romagnani, S., 2009. Heterogeneity of human effector CD4+ T cells. Arthritis Res. Ther. 11, 257.

Arango Duque, G., Descoteaux, A., 2014. Macrophage cytokines: involvement in immunity and infectious diseases. Front. Immunol. 7, 491.

Bodey, G.P., 2000. Unusual presentations of infection in neutropenic patients. Int. J. Antimicrob. Agents 16, 93–95.

Bousfiha, A., Jeddane, L., Al-Herz, W., et al., 2015. The 2015 IUIS phenotypic classification for primary immunodeficiencies. J. Clin. Immunol. 35 (8), 727–738.

Chang, C.-J., Lin, C.-S., Marte, J., Ojcius, D.M., Lai, W.-F., Lu, C.-C., Ko, Y.-F., Young, J.D., Lai, H.-C., 2016. Modulation of host immune response by *Bacteroides fragilis* polysaccharides: a review of recent observations. J. Biomed. Lab. Sci. 28, 11–17.

Cho, J.S., Yi, G., Ramos, R.I., Hebroni, F., Plaisier, S.B., Xuan, C., Granick, J.L., Matsushima, H., Takashima, A., Iwakura, Y., Cheung, A.L., Cheng, G., Lee, D.J., Simon, S.I., Miller, L.S., 2012. Neutrophil-derived IL-1? Is sufficient for abscess formation in immunity against Staphylococcus aureus in mice. PLoS Pathog. 8, e1003047.

Corrales, L., Matson, V., Flood, B., Spranger, S., Gajewski, T.F., 2016. Innate immune signaling and regulation in cancer immunotherapy. Cell Res. 16.

Corthay, A., 2009. How do regulatory T cells work? Scand. J. Immunol. 70, 326–336.

DeWitt, C.A., Bishop, A.B., Buescher, L.S., Stone, S.P., 2006. Hyperimmunoglobulin E syndrome: two cases and a review of the literature. J. Am. Acad. Dermatol. 54, 855–865.

Duan, J., Avci, F.Y., Kasper, D.K., 2008. Microbial carbohydrate depolymerization by antigen-presenting cells: deamination prior to presentation by the MHCII pathway. Proc. Natl. Acad. Sci. U. S. A. 105, 5183–5188.

Glocker, E.-O., Kotlarz, D., Klein, C., Shah, N., Grimbacher, B., 2011. IL-10 and IL-10 receptor defects in humans. Ann. N. Y. Acad. Sci. 1246 (1), 102–107.

Grace, R.H., 1990. The management of acute anorectal sepsis. Ann. R. Coll. Surg. Engl. 72, 160.

Hilchie, A.L., Wuerth, K., Hancock, R.E., 2013. From immune modulation by multifaceted cationic host defense (antimicrobial) peptides. Nat. Chem. Biol. 9, 761–768.

Karaca, N.E., Aksu, G., Ulusoy, E., Aksoylar, S., Gozmen, S., Gene, F., Akarcan, S., Gulez, N., Hirschmug, T., Kansoy, S., Boztug, K., Kutukculer, N., 2016. Early diagnosis and hematopoietic stem cell transplantation for IL10R deficiency leading to very early-onset inflammatory bowel disease are essential in familial cases. Case Reports Immunol., 1–5.

Loddo, I., Romano, C., 2015. Inflammatory bowel disease: genetics, epigenetics, and pathogenesis. Front. Immunol. 6, 551.

Malech, H.L., Hickstein, D.D., 2007. Genetics, biology and clinical management of myeloid cell primary immune deficiencies: chronic granulomatous disease and leukocyte adhesion deficiency. Curr. Opin. Hematol. 14, 29–36.

Malik, A.I., Nelson, R.L., Tou, S., 2010. Incision and drainage of perianal abscess with or without treatment of anal fistula. Cochrane Database Syst. Rev. 7, CD006827.

Martinez, F.O., Gordon, S., 2014. The M1 and M2 paradigm of macrophage activation: time for reassessment. F1000Prime Rep. 6, 13.

Mazmanian, S.K., et al., 2005. An immunomodulatory molecule of symbiotic bacteria directs maturation of the host immune system. Cell 122, 107–118.

Nada, E.M., Alshoaiby, A.N., Alaskar, A.S., Khan, A.N., 2007. Severe anal and abdominal pain due to typhlitis in a leukemic patient. J. Pain Symptom Manag. 34, 457–459.

Nisbet, A.A., Thompson, M.I., 2002. Impact of diabetes mellitus on the presentation and outcomes of Fournier's gangrene. Urology 60 (5), 775–779.

Notarangelo, L., Casanova, J.L., Conley, M.E., et al., 2006. Primary immunodeficiency diseases: an update from the International Union of Immunological Societies primary immunodeficiency diseases classification committee meeting in Budapest 2005. J. Allergy Clin. Immunol. 117, 883–896.

Parks, A.G., 1961. Pathogenesis and treatment of fistula-in-ano. Br. Med. J. 1, 463–469.

Pozzilli, P., Leslie, R.D.G., 1994. Infections and diabetes: mechanisms and prospects for prevention. Diabet. Med. 11, 935–941.

Rogy, M.A., Beinhauer, B.G., Reinisch, W., Huang, L., Pokieser, P., 2004. Transfer of Interleukin-4 and Interleukin-10 in patients with severe inflammatory bowel disease of the rectum. Hum. Gene Ther. 11, 1731–1741.

Rosenzweig, S.D., Holland, S.M., 2004. Phagocyte immunodeficiencies and their infections. J. Allergy Clin. Immunol. 113, 620–626.

Sands, B.E., Anderson, F.H., Bernstein, C.N., et al., 2004. Infliximab maintenance therapy for fistulizing Crohn's disease. N. Engl. J. Med. 350, 876–885.

Scholtzova, H., Do, E., Dhakal, S., Sun, Y., Liu, S., Mehta, P.D., Wisniewski, T., 2016. Innate immunity stimulation via toll-like receptor 9 ameliorates vascular amyloid pathology in Tg-SwDI mice with associated cognitive benefits. J. Neurosci. 15. 1967–16.

Smink, M., Lotgering, F.K., Albers, L., de Jong, D.J., 2011. Effect of childbirth on the course of Crohn's disease; results from a retrospective cohort study in the Netherlands. BMC Gastroenterol. 11. article 6.

Sugrue, J., Nordenstam, J., Abcarian, H., Bartholomew, A., Schwartz, J.L., Mellgren, A., Tozer, P.J., 2017. Pathogenesis and persistence of cryptoglandular anal fistula: a systematic review. Tech. Coloproctol. 21, 425–432.

Surana, N.K., Kasper, D.L., 2012. The yin yang of bacterial polysaccharides: lessons learned from B. Fragilis PSA. Immunol. Rev. 245, 13–26.

Thrasher, A.J., Burns, S., Lorenzi, R., Jones, G.E., 2000. The Wiskott–Aldrich syndrome: disordered actin dynamics in haematopoietic cells. Immunol. Rev. 178, 118–128.

Whiteford, M.H., Kilkenny, J., Hyman, N., Buie, W.D., Cohen, J., Orsay, C., Dunn, G., Perry, W.B., Ellis, C.N., Rakinic, J., Gregorcyk, S., Shellito, P., Nelson, R., Tjandra, J.J., Newstead, G., The Standards Practice Task Force, The American Society of Colon and Rectal Surgeons, 2005. Practice parameters for the treatment of perianal abscess and fistula-in-ano. Dis. Colon Rectum 48, 1337–1342.

Wiese-Posselt, M., Heuck, D., Draeger, A., et al., 2007. Successful termination of a furunculosis outbreak due to lukS-lukF-positive, methicillinsusceptible Staphylococcus aureus in a German village by stringent decolonization, 2002–2005. Clin. Infect. Dis. 44, e88–e95.

Wood, P., Stanworth, S., Burton, J., et al., 2007. Recognition, clinical diagnosis and management of patients with primary antibody deficiencies: a systematic review. Clin. Exp. Immunol. 149, 410–423.

CHAPTER 5

Diagnosis of Perianal Diseases

Riyadh Mohammad Hasan

Contents

1 INTRODUCTION

Anorectal disorders are common, and their prevalence in the general population is probably much higher than that seen in clinical practice and hospitals, as most patients do not seek medical attention because of embarrassment (Gopal, 2002). These disorders affect men and women of all ages. The symptoms of anorectal disorders are nonspecific, ranging from mildly irritating pruritus ani to potentially severe pain (Lacy and Weiser, 2009). The diagnosis of anorectal disorders consists of taking a careful history and performing a physical examination before the patient can be subjected to various forms of investigation.

New Concepts in the Management of Septic Perianal Conditions
https://doi.org/10.1016/B978-0-12-816111-1.00005-7

2 HISTORY

Diagnosis of anorectal disorders begins with a medical history. Pain, bleeding, discharges (either mucoid, purulent or fecal) or change in bowel habits are the common presenting symptoms. It is also important to enquire about other associated illnesses, medications, family history, bleeding tendency, and sexual contacts (Nelson and Cima, 2007). The most evident symptom of perianal abscess is severe pain in the anal region of a short duration, usually 2–3 days. Pain is severe, throbbing in character, and increased by sitting and walking. The differential diagnoses of rectal pain are anal fissure, thrombosed hemorrhoid, levator ani syndrome, proctalgia fugax, coccydynia, fecal impaction, neoplasm (whether rectal, pelvic, or cauda equine), idiopathic, inflammatory bowel diseases (ulcerative colitis, Crohn's disease), solitary rectal ulcer, pruritis ani, trauma, anal sex, constipation, diarrhea, familial rectal pain, endometriosis, pelvic inflammatory disease, prostatitis, and foreign body (Schubert et al., 2009). Other patients may give a history of discharge, which is generally a presenting symptom of fistula-in-ano; the discharge may either be mucoid, purulent, or fecal, and it must be differentiated from that of pilonidal sinus, hidradenitis suppurativa, and infected ruptured sebaceous cyst. Fistula-in-ano is more common in males than females, and usually arises in their third, fourth, and fifth decades of life for unknown reasons. Patients usually have a recurrent leakage of blood and pus after either surgical or spontaneous drainage of an abscess, which is a feature strongly suggestive of the development of a fistula. In some patients, the recurrence of perianal abscess may indicate the presence of an underlying fistula behind.

Other general symptoms like fever (or in advanced cases, rigors and sweating) may be present. Retention of urine may be the presenting complaint in perianal abscesses.

3 PHYSICAL EXAMINATION

The patient is usually examined in left lateral position with buttocks projecting slightly beyond the edge of the table. The perianal area should be inspected for any indurated, hot, tender mass felt in the region, skin tags, excoriations, scars, or any change in color or appearance. It is fairly easy to diagnose an abscess just by looking at and touching it. Intersphincteric abscesses produce few or no external features, but they can be diagnosed by the intense pain they produce, which is usually described as deep inside, and is made worse by coughing and sneezing. In supralevator abscess, the patient presents with fever and deep rectal pain, and there will be minimal features on external examination. Digital examination may show a tender supralevator induration above the anorectal ring. One must pay attention to the presence of fistula, which is a tract connecting two openings, one internal (which is considered the primary opening) and the other external (which is considered the secondary opening). The internal opening is often found at the dentate line, mostly at a crypt in the midline posteriorly. The treating surgeon needs to know critically important features of fistulas in order to be able to deal with them effectively. These are the primary opening, the site(s) of the secondary opening(s), the course of the fistulous track, the presence of secondary extensions, and the presence of associated diseases. The secondary opening can usually be seen as an elevated nodule of granulation tissue near the anal margin exuding pus or serosanguinous fluid. Intersphincteric fistulas usually have an external opening very close to the anal verge. Sometimes, the tract can be felt as an indurated cord, especially in simple superficial fistulas. An external opening that exudes feces or gases is mostly connected to the rectum, not to the anal canal. Defining the internal opening is the most crucial and sometimes difficult part of diagnosis and treatment. In fact, the internal opening is the crypt where the abscess originated. It may be felt during rectal examination as an indurated area. Probing was once used to define the internal opening, but probing in the awake patient is painful and may be dangerous because it may force a false tract. Many methods are used to find this opening in case it can't be felt on rectal examination. Goodsall's rule is a useful method of predicting the site of this opening.

4 DIGITAL RECTAL EXAMINATION

This is very important examination in anorectal diseases and should not be missed. The index finger is lubricated with xylocaine jelly for digital examination, which helps in appreciating any mass, induration, stricture, apart from assessing the resting tone and strength of squeeze pressure. A gloved, lubricated finger is placed at the anal verge and gently inserted through the anal canal into the rectum. Rectal mucosa is systemically examined for benign or malignant lesions. It is possible to feel at least 10 cm from anal verge. Assessment of the anal sphincter is also made with assessment of resting tone and voluntary contraction. In males, the prostate can be assessed whereas in females, the rectocele can be detected after pushing the vaginal wall forward. (Nelson and Cima, 2007). Digital examination is probably the best way of identifying the swelling, but pain may make this examination too difficult or even impossible. Sometimes, the patient may need examination under anesthesia to get an accurate diagnosis.

5 ANOSCOPY/PROCTOSCOPY

A visual examination of the lower part of rectum and anal canal through a proctoscope must be done. Anoscopy and proctosigmoidoscopy can detect any growth in the anal canal. Anoscopy enables a satisfactory examination of anal canal and distal rectum. Endoscopy in the form of sigmoidoscopy or proctoscopy is used for detecting the primary opining of the fistula but usually an anoscope can suffice, which may show the primary opening as an enlarged papilla. For complete examination of the anorectum, proctosig-moidoscopy is the preferred method. Any suspicious area can be biopsied. The proctoscope with obturator in situ is well lubricated and introduced into the anal canal; while introducing the device, one must remember that the anal canal is directed upwards and forwards towards the umbilicus of the patient. After it has been fully introduced, the obturator is taken out, and the inside of the proctoscope is well illuminated. Hemorrhoids, internal opening of the fistulous tract, anal polyps, fissures, and ulcerations can be identified (Garg et al., 2011). If doubt still exists about the location of the primary opening, a trial of injection of hydrogen peroxide with or without methy-lene blue and visualization of bubbles at the internal opening either directly or through endoscopy or transrectal ultrasonography. This method is also useful in the operating theater to define the internal opening prior to prob-ing the fistula. A sigmoidoscope has the additional benefit of ruling out the presence of inflammatory bowel disease, diverticulitis with perforation and fistulization, and more seriously, low rectal and anal canal carcinoma presenting with fistula.

6 LABORATORY INVESTIGATIONS

6.1 Stool Examination

Stool Examination if infectious diarrhea or sexually transmitted disease is suspected.

6.2 Histopathological Examination

Histopathological Examination is done to know the histological diagnosis of the mass or the suspicious area seen in proctosigmoidoscopy.

7 IMAGING STUDIES

Some rare deep abscesses might require ultrasound, CT scan, or even MRI for an accurate diagnosis. Imaging studies which can help determine the diagnosis in cases of a deep nonpalpable perirectal abscess include pelvic CT scan, MRI or trans-rectal ultrasound. These studies are not necessary for cases in which the diagnosis can be made upon physical examination (Beets-Tan et al., 2001).

7.1 Fistulography

Fistulography is a good diagnostic method for detecting the internal opening in a fistula-in-ano, and it is still used in spite of the introduction of newer, more informative methods. This simple and affordable procedure can be done in an outpatient clinic, but it has two weak points: the inability to define secondary extensions due to insufficient filling with contrast, and the nonvisualization of anal sphincters, which is the most important factor determining the outcome of continence (Halligan and Stoker, 2006).

7.2 Anorectal Endosonography (Endorectal Ultrasound)

This was the first method to show the details of anal wall anatomy, especially the sphincters (Law and Bartram, 1989). It can be used with or without injection of hydrogen peroxide, but it has an unfortunately limited field of view, making it excellent imaging procedure for the diagnosis of inter-sphincteric fistulas and their relationship to the anal sphincters, but less so for primary superficial, suprasphincteric, and extrasphincteric tracks or secondary extensions (Van Outryve et al., 1991, 1994).

These limitations may be overcome by the improved US transducers with three-dimensional (3D) equipment (Saranovic et al., 2007). Currently, transrectal ultrasound is reserved for complicated cases, especially recurrent fistulas and supralevator fistulas where it is important to show the fistula's relationship to the sphincters, and there is hope that it may be more widely used in the future. Transrectal ultrasounds are an accurate means of delineating anatomy in relation to a fistula. It is easily performed and less expensive than Magnetic Resonance Imaging (MRI), but it is not appropriate for the patient with severe anal pain or an anatomical stricture. The exact choice of which modality to use depends on local expertise, cost, and the equipment available.

7.3 Computed Tomography

Computed tomography (CT) is a useful technique for extrasphincteric fistulas, but it has limitations. It is used with rectal and intravenous contrast material, but it has poor resolution for soft tissue, which makes it unable to analyze anal fistulous tracks (Guillaumin et al., 1986; Yousem et al., 1988). The overall sensitivity of computed tomography in identifying abscess is 77%, and it lacks sensitivity in detecting perirectal abscess, particularly in the immunocompromised patient (Caliste et al., 2011).

7.3.1 Virtual Colonoscopy

This procedure is done using helical CT; contrast agents may be used orally or rectally with insufflations. The accuracy of this technique may approach that of colonoscopy.

7.4 MRI (Magnetic Resonance Imaging)

Adopting endoanal coils and phased array imaging has contributed to the evolution of using MRI to evaluate anorectal disease (Buchanan et al., 2004; (Berman et al., 2007) (Fig. 1). Magnetic resonance imaging is now considered the "gold standard" for fistula imaging, but it is limited by its availability and cost and is usually reserved for difficult, recurrent cases. It is used in the detection of secondary extensions, which are one of the main causes of surgical failure. In addition, it has the ability to study the integrity of the sphincter complex and to define the relationship of the fistulous track to the anal sphincters before performing fistulotomy to be sure of preserving continence. (Spencer et al., 1998; Chapple et al., 2000).

In cases of recurrence and severe inflammation, the clinical examination will have difficulty delineating the position of the abscess and/or fistulous tract making planning surgical intervention very difficult and the risk of surgical failure higher. Here, MRI may indicate the site of abscess in cases in which localization is difficult, that is, intersphincteric and supralevator abscesses (Fig. 2). MRI may show the extension of an existing abscess through both ischioanal fossae to form a horseshoe abscess (Fig. 3). MRI is also useful for showing the external and internal openings and the track, and it is especially useful in recurrent fistula and cases in which there are secondary extensions (Figs. 4 and 5).

MRI is useful in monitoring therapy, especially in complicated fistulas and in Crohn's disease fistulas, and when using fibrin glue, because sometimes the external opening is closed, but the track is still not healed.

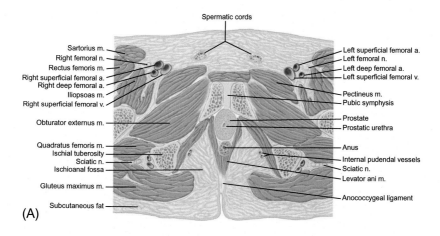

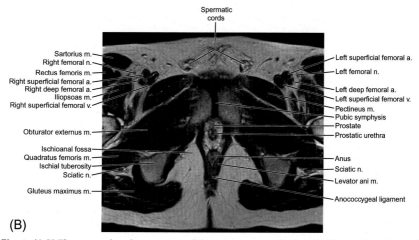

Fig. 1 (A,B) The normal male anatomy of the perineum at the level of the mid anal canal in the axial plane. ES = external sphincter, IA = ischioanal fossa, InS = intersphincteric space, IS = internal sphincter. *(From Torigian, D.A., Hammell, M.T.K., 2013. Netter's Correlative Imaging: Abdominal and Pelvic Anatomy. Elsevier-Saunders, pp. 268–269 (Chapter 6).)*

MRI is excellent for showing the extent of healing and the effect of infliximab or fibrin glue in the process of healing (Buchanan et al., 2003; Van Assche et al., 2003). MRI may also guide surgical treatment, not only preoperatively but also intraoperatively to identify the extensions of the fistula, preventing incomplete procedures and the possibility of recurrence, especially in complex and recurrent fistulas (Gould et al., 2002). Buchanan et al. recently studied a cohort of 108 patients with recurrent fistulas using digital examination, endoanal ultrasound, and MRI on each patient. Digital

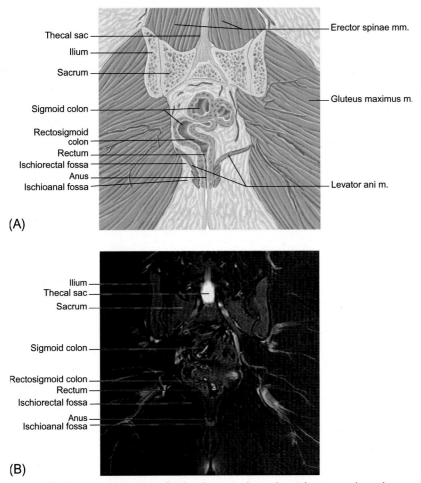

Fig. 2 (A,B) High signal intensity fluid collection along the right posterolateral aspect of the anal canal with Simple intersphincteric fistula. *(Edited from Torigian, D.A., Hammell, M.T.K., 2013. Netter's Correlative Imaging: Abdominal and Pelvic Anatomy. Elsevier-Saunders, pp. 474–475 (Chapter 5).)*

examination correctly identified 61% of tracks, compared with 81% of tracks found by endoanal ultrasound, and 90% by MRI (Buchanan et al., 2004).

8 MANOMETRY

Surgeons may request anal manometry to further investigate problems of incontinence in defecation. Manometry tests the strength of sphincter muscle. Some authors advocate preoperative manometry in order to choose the

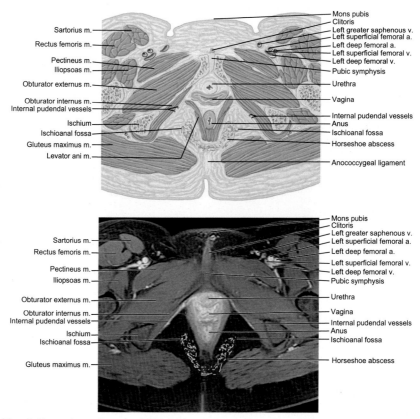

Fig. 3 Horseshoe abscess. *(Edited from Torigian, D.A., Hammell, M.T.K., 2013. Netter's Correlative Imaging: Abdominal and Pelvic Anatomy. Elsevier-Saunders, pp. 452 (Chapter 5).)*

correct therapeutic management according to the risk of incontinence (Grade B) (Holzheimer and Mannick, 2001).

9 DEFECOGRAPHY

Barium defecography is performed by injecting barium contrast mixed with Metamucil or another thickening agent into the rectum and taking lateral images of the anorectum during pelvic floor contraction, before, during, and after attempted defecation (Agachan et al., 1996). The angle between the axis of the rectum and the anal canal provides an indirect measure of whether the puborectalis muscle relaxes, which is the normal response, or contracts, indicative of defecatory disorder, during simulated defecation. Extra information is obtained on structural causes of outlet dysfunction

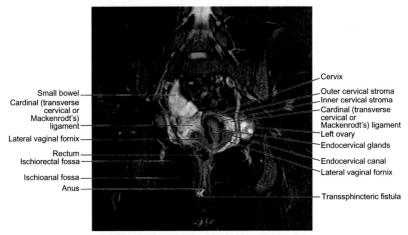

Fig. 4 Transsphincteric fistula is distinguished by location of the internal opening in the middle third of the anal canal. *(Edited from Torigian, D.A., Hammell, M.T.K., 2013. Netter's Correlative Imaging: Abdominal and Pelvic Anatomy. Elsevier-Saunders, pp. 473 (Chapter 5).)*

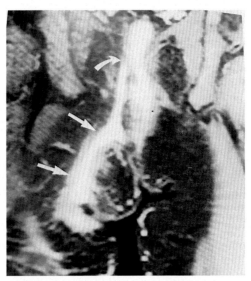

Fig. 5 Coronal MRI STRI image reveals a right sided extrasphincteric fistula (straight arrows) with its enteric communication in the rectum (curved arrow). *(From Chapman, A.H., Guthrie, J.A., Robinson, J.A., 2003. The stomach and duodenum. In: Sutton, D. (Ed.), Text Book of Radiology and Imaging, vol. 1, seventh edition. Churchill Livingstone, pp. 654 (Chapter 19).)*

including rectal prolapse, rectocele, and enterocele. Defecography was once regarded as the gold standard for diagnosis of defecatory disorder. Consequently, defecography can identify impaired evacuation in patients with symptoms of defecatory disorder, but with normal electromyography testing (Chiarioni et al., 2014).

10 ELECTROMYOGRAPHY

Electromyography (EMG) may help in the assessment of anorectal incontinence, constipation, or any other pelvic floor disorder (Nelson and Cima, 2007).

Anorectal disease is common and its incidence is increasing over the last few decades among populations. Diseases of the ano-rectum are usually easy to diagnose via patient's history, which provides a great deal of information, and clinical examination with digital examination supplemented with anoscopy or proctoscopy help in diagnosis. Specialized investigations are needed in selected group of patients (Garg et al., 2011).

REFERENCES

Agachan, F., Pfeifer, J., Wexner, S.D., 1996. Defecography and proctography. Results of 744 patients. Dis. Colon Rectum 39, 899–905.
Beets-Tan, R.G., Beets, G.L., van der Hoop, A.G., Kessels, A.G., Vliegen, R.F., Baeten, C.G., van Engelshoven, J.M., 2001. Preoperative MR imaging of anal fistulas: does it really help the surgeon? Radiology 218, 75–84.
Berman, L., Israel, G.M., McCarthy, S.M., Weinreb, J.C., Longo, W.E., 2007. Utility of magnetic resonance imaging in anorectal disease. World J. Gastroenterol. 13, 3153–3158.
Buchanan, G.N., Bartram, C.I., Phillips, R.K., et al., 2003. Efficacy of fibrin sealant in the management of complex anal fistula: a prospective trial. Dis. Colon Rectum 46, 1167–1174.
Buchanan, G.N., Halligan, S., Bartram, C.I., Williams, A.B., Tarroni, D., Cohen, C.R., 2004. Clinical examination, endosonography, and MR imaging in preoperative assessment of fistula in ano: comparison with outcome-based reference standard. Radiology 233, 674–681.
Caliste, X., Nazir, S., Goode, T., et al., 2011. Sensitivity of computed tomography in detection of perirectal abscess. Am. Surg. 77, 166–168.
Chapple, K.S., Spencer, J.A., Windsor, A.C., Wilson, D., Ward, J., Ambrose, N.S., 2000. Prognostic value of magnetic resonance imaging in the management of fistula-in-ano. Dis. Colon Rectum 43, 511–516.
Chiarioni, G., Kim, S.M., Vantini, I., et al., 2014. Validation of the balloon evacuation test: reproducibility and agreement with findings from anorectal manometry and electromyography. Clin. Gastroenterol. Hepatol. 12, 2049–2054.
Garg, H., Singh, S., Bal, K., 2011. Approach to the diagnosis of anorectal disorders. JIMSA 24, 89–90.

Gopal, D.V., 2002. Diseases of the rectum and anus: a clinical approach to common disorders. Clin. Cornerstone 4, 34–48.

Gould, S.W., Martin, S., Agarwal, T., Patel, B., Gedroyc, W., Darzi, A., 2002. Image-guided surgery for anal fistula in a 0.5T interventional MRI unit. J. Magn. Reson. Imaging 16, 267–276.

Guillaumin, E., Jeffrey, R.B., Shea, W.J., Asling, C.W., Goldberg, H.I., 1986. Perirectal inflammatory disease: CT findings. Radiology 161, 153–157.

Halligan, S., Stoker, J., 2006. Imaging of fistula in ano. Radiology 239, 18–33.

Holzheimer, R.G., Mannick, J.A. (Eds.), 2001. Surgical Treatment: Evidence-Based and Problem-Oriented. Zuckschwerdt, Munich.

Lacy, B.E., Weiser, K., 2009. Common anorectal disorders: diagnosis and treatment. Curr. Gastroenterol. Rep. 11, 413–419.

Law, P.J., Bartram, C.I., 1989. Anal endosonography: technique and normal anatomy. Gastrointest. Radiol. 14, 349–353.

Nelson, H., Cima, R.R., 2007. Anus. In: Sabiston Textbook of Surgery, eighteenth ed,.

Saranovic, D., Barisic, G., Krivokapic, Z., Masulovic, D., Djuric-Stefanovic, A., 2007. Endoanal ultrasound evaluation of anorectal diseases and disorders: technique, indications, results and limitations. Eur. J. Radiol. 61, 480–489.

Schubert, M.C., et al., 2009. What every gastroenterologist needs to know about common anorectal disorders. World J. Gastroenterol. 15, 3201–3209.

Spencer, J.A., Chapple, K., Wilson, D., Ward, J., Windsor, A.C., Ambrose, N.S., 1998. Outcome after surgery for perianal fistula: predictive value of MR imaging. AJR Am. J. Roentgenol. 171, 403–406.

Torigian, D.A., Hammell, M.T.K., 2013. Netter's Correlative Imaging: Abdominal and Pelvic Anatomy. Elsevier-Saunders, pp. 268–269, 452, (Chapters 5 and 6).

Van Assche, G., Vanbeckevoort, D., Bielen, D., et al., 2003. Magnetic resonance imaging of the effects of infliximab on perianal fistulizing Crohn's disease. Am. J. Gastroenterol. 98, 332–339.

Van Outryve, M.J., Pelckmans, P.A., Michielsen, P.P., Van Maercke, Y.M., 1991. Value of transrectal ultrasonography in Crohn's disease. Gastroenterology 101, 1171–1177.

Van Outryve, M., Pelckmans, P., Fierens, H., Van Maercke, Y., 1994. Transrectal ultrasonographic examination of the anal sphincter. Acta Gastroenterol. Belg. 57, 26–27.

Yousem, D.M., Fishman, E.K., Jones, B., 1988. Crohn disease: perirectal and perianal findings at CT. Radiology 167, 331–334.

FURTHER READING

SWCRS (n.d.) http://www.swcrs.com.au/.

CHAPTER 6

Management of Perianal Suppuration

Riyadh Mohammad Hasan

Contents

1 INTRODUCTION

Anorectal abscesses and anal fistulas represent the same disease process and are two of the most common problems seen at different times in hospitals and clinics. Ano–rectal complaints are usually benign in origin. Most patients suffering from these disorders usually do not seek medical advice at an early stage due to embarrassment. This results in progression of the disease and significant disturbance to the quality of life. Abscess is the acute manifestation of the disease, whereas fistula is the chronic phase of anorectal suppuration. Because of the close relationship of abscess and fistula in their etiology,

New Concepts in the Management of Septic Perianal Conditions
https://doi.org/10.1016/B978-0-12-816111-1.00006-9
45

epidemiology, anatomy, pathophysiology, management, and morbidity, it is appropriate to consider them as one disease abscess-fistula (Gordon, 1992).

2 PERIANAL ABSCESS

2.1 Clinical Presentation/Examination

With anorectal abscess, the patient's main symptoms are fever, toxicity, and pain. This pain must be differentiated from other causes of anal pain like anal fissure, thrombosed hemorrhoids, levator ani spasm, sexually transmitted diseases that affect the anal canal, proctitis, and anal cancer (Ward et al., 1987). The character of this pain varies from minimal discomfort to severe, debilitating pain exacerbated by movement (Calata et al., 2017). On examination, there is tender, erythematous cellulitis and fluctuant swelling due to accumulation of pus that dissects inferiorly in the intersphincteric space reaching the anal verge. When pus traverses the external sphincter into the ischioanal fossa, the abscess tends to be larger, and fewer cutaneous findings are observed. Pus can extend on one or both ischioanal fossa, to form a horseshoe-shaped collection of pus, or it may track up towards and through the levator ani muscles (Whiteford et al., 2005). Intersphincteric and supralevator abscess may be associated with urinary symptoms, but with less obvious external findings, and it is often impossible to perform a careful examination in this condition. A patient with a supralevator abscess sometimes complains of gluteal pain with a tender mass palpated on rectal or vaginal examination (Goldberg et al., 1980a). The presence of a black spot during examination may be indicative of a widespread necrotizing fasciitis (Bubrick and Hitchcock, 1979). Sometimes the pain is so severe, coupled with the patient's resistance, examination must be performed under anesthesia. Imaging is rarely needed to diagnose an acute perianal abscess (Abcarian, 2011).

2.2 Objectives in Management of Perianal Abscess

The objective of surgical therapy for anorectal sepsis is primarily surgical; the classical teaching is that abscess anywhere should be treated with adequate drainage as soon as possible. Drain any associated sepsis in adjacent anatomic spaces, identify a fistula tract, and if found, either proceed with primary fistulotomy to prevent recurrence or leave management of the fistula to a later stage. Pus is sent for microbiological culture, and tissue from the wall of the abscess is sent for histological examination when suspicion arises to exclude specific causes for the abscess (Grace, 1990).

2.3 Treatment of an Acute Anorectal Abscess

Treatment is usually done under anesthesia, and the type of anesthesia used depends on the type of the abscess. In perianal abscess, local anesthesia in the outpatient clinic is sufficient due to the superficial location of the abscess. However, in the case of supralevator abscesses, deep intersphincteric abscesses, big, tense, ischioanal abscesses, or in immunocompromised and diabetic patients, general anesthesia may be needed in an operating theater (Schwartz et al., 2001).

The incision used in the drainage of the abscess is thought to be most beneficial if it is a cruciate incision with trimming of the corners or elliptical in shape (Read and Abcarian, 1979; Kronborg and Olsen, 1984). These incisions are thought to be important in delaying premature closure of the abscess draining site. The incision is made over the most prominent or fluctuant part of the abscess involving the skin and subcutaneous tissue (Corman, 1993) as shown in Fig. 1.

The abscess cavity is curetted and irrigated thoroughly, because inadequate drainage may result in recurrence or, especially in immunocompromised patients, may result in necrotizing soft tissue infection of the perineum. Although probing is very difficult and probably hazardous during the acute abscess stage, in experienced hands, probing may find an internal fistulous opening which could be the causative factor for the abscess. If probing was successful in finding an internal opening, a seton thread is passed

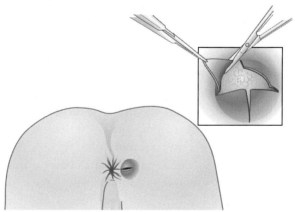

Fig. 1 Diagram illustrate the drainage of abscess. Cruciate incision and excision of skin and drainage cavity. *(Nelson, H., Cima, R.R., 2008. Anus. In: Townsend, C.M., Beauchamp, R.D., Evers, B.M., Mattox, K.L. (Eds.), Sabiston Text Book of Surgery, 18th ed. Saunders-Elsevier, pp. 1446 (Chapter 51).)*

through to enhance drainage and assist in future healing. Immediate fistulotomy during this stage is best avoided (Malik et al., 2010; Lohsiriwat, 2016).

2.4 Types of Anorectal Abscess

2.4.1 Ischioanal Abscess

The swelling will be most prominent over the ischioanal fossa, and drainage is similar to perianal abscess in regard to the site of incision, which should be over the most prominent and fluctuant area of the abscess. The cavity should be dealt with by curetting and irrigation (Hopping, 1956). If the ischioanal abscess is bilateral, then the site of origin is in the deep postanal space, and here, the incision should be a longitudinal one between the tip of the coccyx and anus, deeper into the anococcygeal ligament to enter the deep postanal space. Again, the abscess cavity should be dealt with by curetting and irrigation (Lohsiriwat et al., 2010).

2.4.2 Intersphinteric Abscess

Perianal and ischioanal abscesses are characterized by apparent swelling and induration, unlike intersphinteric abscesses, in which these signs are minimal. In the case of intersphinteric abscesses, the abscess is suspected because of extreme pain which may prevent completion of rectal examination. Most of these abscesses are located in the posterior quadrant and are usually caused by fistulization from the posterior anal canal, usually in the bed of a chronic posterior anal fissure. During rectal examination, it is felt to be an indurated or bulging mass above the dentate line. Here, drainage is made through the anal canal deep to the internal sphincter, followed by curetting and irrigation. No packing or drain is needed (Ommer et al., 2017). Low intersphincteric abscesses should be treated by de-roofing of the abscess and division of the internal sphincter up to the level of the dentate line. High intersphincteric abscesses are relatively frequent and mostly require staged surgery with a temporary mushroom (de Pezzer) catheter. The incision should be placed near the anus over the most fluctuating part of the abscess to decrease the tissue that must be opened if a fistula is found (Fig. 2).

Then, a 10–16 French soft latex mushroom catheter is inserted over a probe into the abscess cavity, and the size and length of the catheter should correspond to the size of the abscess cavity, neither too short nor too long (Fig. 3).

The catheter tip will hold the catheter in place in the abscess, and the external portion of the catheter is shortened to leave 2–3 cm outside the skin, with the tip in the depth of the abscess cavity (Fig. 4).

Fig. 2 Catheter drainage of an abscess via a stab incision near the anus. *(Edited from Nelson, H., Cima, R.R., 2008. Anus. In: Townsend, C.M., Beauchamp, R.D., Evers, B.M., Mattox, K.L. (Eds.), Sabiston Text Book of Surgery, 18th ed. Saunders-Elsevier, pp. 1446 (Chapter 51).)*

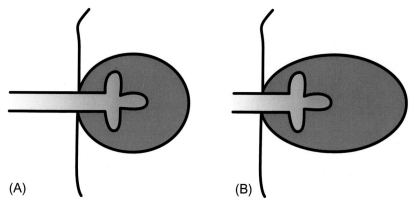

(A)   (B)

Fig. 3 Catheter in an abscess cavity. (A) Correct size and length of catheter. (B) Catheter too short.

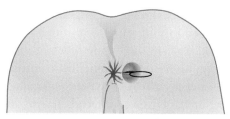

Fig. 4 Catheter drainage in an abscess cavity. *(Edited from Nelson, H., Cima, R.R., 2008. Anus. In: Townsend, C.M., Beauchamp, R.D., Evers, B.M., Mattox, K.L. (Eds.), Sabiston Text Book of Surgery, 18th ed. Saunders-Elsevier, pp. 1446 (Chapter 51).)*

One must judge the size of the abscess, the amount of granulation tissue around the catheter, and the character and amount of drainage from the abscess. A small bandage is placed over the catheter.

Accurate anatomical ultrasound localization and proper drainage become important to avoid recurrences or formation of extrasphincteric fistulas (Millan et al., 2006). Sepsis in intersphincteric space has an important role in pathogenesis of fistula-in-ano. This sepsis is like a small abscess in a closed space. This closed space needs to be drained adequately and then kept open to heal properly. The new approach was to open and drain the intersphincteric space through internal opening via transanal approach. Transanal laying open of the intersphincteric space (TROPIS) was done through an internal opening. The external sphincter was not cut. The tracts in the ischioanal fossa were curetted and cleaned. TROPIS is a simple, effective sphincter-sparing procedure to treat highly complex fistula-in-ano including supralevator and horseshoe fistula, and this was done in submucosal and intersphincteric anorectal abscesses (Garg, 2017a, b, c).

2.4.3 Supralevator Abscess

This may present as an acute intraabdominal condition. It is uncommon and difficult to diagnose. Rectal examination reveals tender bulging on either side of the rectum or posteriorly above the anorectal ring. It can arise in different ways (see pages on pathogenesis). Drainage should be done after determining its origin. If it has originated from upward extension of intersphincteric abscess, it should be drained to the rectum, not to the ischioanal fossa, or a complicated suprasphincteric fistula will develop (Fig. 5).

If the abscess has originated from an upward extension of ischioanal abscess it should be drained to the ischioanal fossa, drainage to the rectum may result in extrasphincteric fistula. If the abscess is a complication of intraabdominal disease (appendiceal, gynecological, diverticular disease, Crohn's disease, or malignancy), then the original disease should be treated, and the abscess is drained to the rectum or through the ischioanal fossa or through the abdominal wall. In some immunocompromised patients with severe sepsis, sometimes we need a defunctioning colostomy with the drainage procedure (Ortega et al., 2015). The new technique was used to treat supralevator abscesses upward from an intersphincteric origin in two stages; first, an endoanal drainage was performed by inserting a mushroom catheter into the supralevator abscess cavity. In the second stage, transanal unroofing of the abscess was performed with an endostapler. So the use of an endostapler in the treatment of supralevator abscess of intersphincteric origin may be

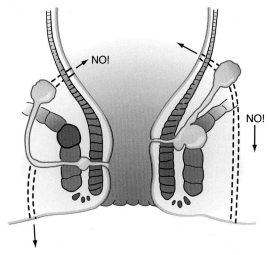

Fig. 5 Correct drainage routes for supralevator abscesses. *(From Nelson, H., Cima, R.R., 2008. Anus. In: Townsend, C.M., Beauchamp, R.D., Evers, B.M., Mattox, K.L. (Eds.), Sabiston Text Book of Surgery, 18th ed. Saunders-Elsevier, pp. 1446 (Chapter 51).)*

an alternative method to decrease the risk of recurrence and anal incontinence (García-Granero et al., 2014). Magnetic resonance imaging seems essential to clarify the perfect location of supralevator abscess, its origin, and choice of the right drainage route (Garcia-Granero et al., 2014).

In addition to incision and drainage, surgeons practice the procedure of packing the abscess cavity. The packing is intended to provide hemostasis and to prevent closure of the overlying skin, thus allowing healing by secondary intention (Tonkin et al., 2004). There is currently no established, evidence-based medicine for this practice. On the other hand, regular postoperative packing can be painful and requires multiple visits to a hospital or home visits by a nurse to change the packing until healing has occurred (O'Malley et al., 2009). The cost of home visits by a nurse and the inconvenience to patients prevents them from return to usual activities. Many authorities discourage such practice because they think that this helps to maintain the potential space and because of the extreme pain experienced during changing of the packs (Perera et al., 2015). Those who encourage packing claim that its use minimizes the size of the wound needed, and the cavity is kept open, encouraging drainage of pus, and preventing premature closure of the skin.

Hasan (2016) used postoperative insertion of a corrugated rubber drain in the abscess cavity, fixed to the skin by suture, to prevent slipping into the wound. It has the advantage of keeping the space draining preventing premature closure and recurrence of the abscess; in addition to that, it avoids the extreme pain encountered during packing change, if a pack is used. A track may be kept open by inserting a sheet of latex rubber or plastic material, which is often corrugated to create spaces. Fluid reaches the surface by gravity or vis a tergo (Latin for "push from behind"), where it must be soaked up by gauze packs. To prevent slipping into the wound, these were affixed by stitching them to the skin or by placing a large safety pin through the projecting portion (basic surgical techniques). This drain is simply inserted into the abscess cavity in the case of perianal abscess, or inserted in one or both ischioanal spaces if the abscess was of the horseshoe type (Hasan, 2016).

Large bilateral ischioanal (horseshoe) abscesses require incision and drainage under regional or general anesthetic. The patient is usually placed in the prone jackknife or left lateral (Sims) position. The abscess is usually incised using the Hanley procedure by a midline incision between the coccyx and anus, spreading the superficial external sphincter to enter the space. An opening is made in the posterior midline, and the lower half of the internal sphincter is divided to drain the anal gland in which the infection originated (Goldberg et al., 1980b). Counter-incisions are made over each ischioanal fossa to allow drainage of the anterior extensions of the abscess (Hanley et al., 1976) (Fig. 6).

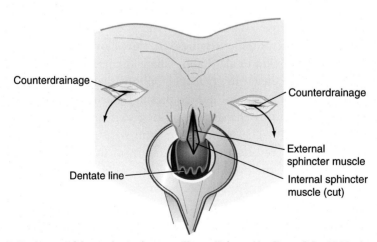

Fig. 6 Drainage of horseshoe abscess. *(From Nelson, H., Cima, R.R., 2008. Anus. In: Townsend, C.M., Beauchamp, R.D., Evers, B.M., Mattox, K.L. (Eds.), Sabiston Text Book of Surgery, 18th ed. Saunders-Elsevier, pp. 1447 (Chapter 51).)*

Other techniques used include *kṣārasūtra* (medicated thread), the WHO-accepted best alternative, surgical procedure to combat complications, which is successfully practiced in ayurvedic colleges and other medical centers throughout India (WHO). *Kṣārasūtra* (Acharya, 1994a, b) was first described in *Suśruta Saṃhitā*, later by *Cakrapāṇidatta* (11 CE) (Dwivedy, 2011). In 1964, the conceptual basis for revival of *kṣārasūtra* preparation was laid down by Dr. Shankaran and Dr. Pathak under the guidance of Prof. Deshpande at the Department of Shalya-Shalakya, PGIIM, BHU, Varanasi. *Kṣārasūtra* is prepared with surgical linen (Barbour) 20 thread coated with latex of *snuhi* (*Euphorbia neriifolia*), *haridrā* (*Curcuma longa*) powder and *kṣāra* made from the whole *apāmārga* plant (*Achyranthes aspera* Linn., *Amaranthaceae*). The patient was kept in a lithotomy position, and the perianal region was cleaned with savlon, spirit, and betadine, consecutively, and draping was done. A local anesthetic of 2% lignocaine with adrenaline diluted with distilled water to 1% dilution was infiltrated. With the help of a number 15 blade, the abscess was incised and drained; after 1 week, the cavity was probed, and primary threading was done and treated like fistula-in-ano. With regard to modes of action, kṣārasūtra is supported by antibacterial theory, as one of its ingredients is turmeric. Turmeric is universally known for its antibacterial, antifungal, and anti-inflammatory properties (Faujdar et al., 1981; Ibn Majdoub Hassani et al., 2012). Chemical cauterization by necrosis of unhealthy granulation and proliferation of new connective tissue takes place under the influence of *kṣāra* present in the thread (which is prepared with *Achyranthes*) and the corrosive nature of latex made with *Euphorbia*. The suture serves as a local drug-delivery system through drugs incorporated into the thread, which are delivered layer by layer to the local pathological tissue planes, and it debrides the unhealthy granulation, enhancing the healing process and through mechanical action, exerting mechanical pressure on the tissue. So, *kṣāra* acts as a powerful debridement agent and selectively acts on unhealthy granulation, pus pockets, etc. This process of debridement and healing starts from deeper tissues and travels towards the periphery (Acharya, 1994a, b). Consequently, the *kṣārasūtra* technique leads to prompt symptomatic resolution and prevention of recurrence and further complications (Kurapati and Nishteswar, 2014).

The problem of antibiotics use is subject to controversy among health providers (Stewart et al., 1985). Some believe that antibiotics are not needed routinely in perianal abscess if the patient is otherwise healthy with regard to immune deficiency, and in those patients with prosthesis or cardiac valvular disease, especially those with Crohn's disease, where metronidazol or ciprofloxacin have superior effects. Other investigators believe that antibiotics should be used with surgery during the period of pre- and postoperative periods (Hasan, 2017), additionally, they think that choice of antibiotics makes a lot of difference in the postoperative sequelae if the antibiotic was of the bacteriostatic or bacteriocidal category. Antibiotics may be chosen according to the result of culture and sensitivity of the pus drained during the drainage procedure. Moreover, the type of bacteria isolated may provide a clue about an underlying fistula, if it was found to be gut-type organisms (Stewart et al., 1985).

3 FISTULA-IN-ANO

3.1 Clinical Presentation/Examination

Anal fistulas continue to be a problem for both patients and surgeons despite scientific advances. The majority of fistulae-in-ano are cryptoglandular, and the minority are nonspecific or idiopathic. Rarely, it may be associated with TB, Crohn's disease, malignancy, lymphogranuloma venereum, and actinomycosis. Fistulas may occur less frequently from trauma or iatrogenic perforation, posthemorrhoidectomy, infected episiotomy, or repair of a fourth-degree sphincter tear during delivery, and infected anal fissure. Recurrent or long-standing fistulas should raise the suspicion of high type of fistula (Chrabot et al., 1983). A full medical, obstetric, gastrointestinal, anal surgical, and continence history must be obtained from the patient. Patients often provide a history of cyclic pain, swelling, and spontaneous or planned surgical drainage of an anorectal abscess. Signs and symptoms of fistula-in-ano are perianal discharge, pain, swelling, bleeding, diarrhea, skin excoriation, and an external opening. Physical findings are the mainstay of diagnosis. Proctosigmoidoscopy is necessary to gain information about sphincter strength and to exclude associated conditions. First, there must be a determination of the site of the internal opening, the site of the external opening, the course of the primary track, the presence of secondary extensions, and the presence of other associated conditions. Palpable induration between external opening and anal margin suggests a relatively superficial track, whereas supralevator induration suggests a primary track above the levators or high in the roof of the ischiorectal fossa, or a high secondary extension. Examination must be done for the entire perineum, external opening of the fistula, and elevation of the granulation tissue, and the site of the internal opening may be felt as a point of induration or seen as an enlarged papilla, and any discharge from an external opening like blood, pus, or fecal material should be noted. Digital rectal examination must be done to detect any fibrous tract or cord beneath the skin. It also helps to delineate any further acute inflammation and lateral or posterior indurations. The examiner should determine the relationship between the anorectal ring and the position of the tract before the patient is relaxed by anesthesia. The sphincter tone and voluntary squeeze pressures should be assessed before any surgical intervention to determine whether preoperative manometry is indicated. Anoscopy is usually required to identify the internal opening. Goodsall's rule is still applicable unless the anatomy has been distorted by prior operations and fibrosis (Goodsall, 1900). Proctoscopy is also indicated in the presence of rectal disease. Most patients cannot tolerate even gentle probing of the fistula tract in the office, and this practice should be avoided (Corman, 2005). Then proctosigmoidoscopy should be done to assess the health state of the anorectal mucosa, and a biopsy should be taken from unhealthy mucosa for histopathology. Long-standing fistulae may be the site of malignancy formation (Belliveau, 1990). Probing the fistula tract in the office is painful and unnecessary. Full examination under anesthesia should be repeated before surgical intervention. If a primary opening cannot be easily identified intraoperatively, injection of dilute hydrogen peroxide solution with or without a few drops of methylene blue is often helpful. This is instilled via the external opening, could be a very useful way of demonstrating the site of the internal opening; gentle use of probes and a finger in the anorectum usually delineates primary and secondary tracks and their relations to the sphincters. If one or all the above methods have not resulted in identification of the internal opening, anal fistulography, endoanal ultrasound with injection of peroxide, computed tomography (CT), or magnetic resonance imaging (MRI) may be utilized (Nelson et al., 1985).

3.2 Objectives in Management of Fistula-in-Ano

The goal of treatment of fistula-in-ano is the eradication of sepsis without sacrificing continence. Surgery is the mainstay of treatment, and its aim is to cure fistulas while preserving the sphincter mechanism and preventing recurrence. Surgical treatment should aim at healing the fistulous track without affecting continence (Williams et al., 2007).

3.3 Treatment of an Acute Fistula-in-Ano

There is no single method suitable for the treatment of all fistulae-in-ano, and, therefore, treatment must be driven by the surgeon's skill and decision. One must pay attention to the extent of operative sphincter division, postoperative healing rates, and functional detriment (Kodner et al., 1993).

Drainage of an anorectal abscess results in cure in about 50% of patients. The remaining 50% develop a persistent fistula–in–ano (Chrabot et al., 1983). Treatment is achieved by primary fistulotomy, which should only be performed in cases of superficial fistulas and by experienced surgeons.

3.4 Types of Acute Fistula-in-Ano

Four main types of fistula are categorized according to the sphincter muscles. They share the property of having an internal opening into the anorectal lumen and external opening to the skin (Parks et al., 1976). They are the following (Abcarian, 2011) (Fig. 7):

1. Intersphincteric Fistula
2. Transsphincteric Fistula
3. Suprasphincteric Fistula
4. Extrasphincteric Fistula

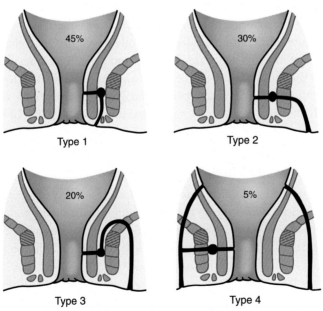

Fig. 7 Diagram of Parks classification of fistula-in-ano. (Type 1) Intersphincteric Fistula, (Type 2) transsphincteric fistula, (Type 3) suprasphincteric fistula, and (Type 4) extrasphincteric fistula. *(From Nelson, H., Cima, R.R., 2008. Anus. In: Townsend, C.M., Beauchamp, R.D., Evers, B.M., Mattox, K.L. (Eds.), Sabiston Text Book of Surgery, 18th ed. Saunders-Elsevier, pp. 1447 (Chapter 51).)*

3.4.1 Intersphincteric Fistula

The intersphincteric or transsphincteric fistula constitutes about 45% of all cases, and do not cross the external sphincter. Instead, they run directly from the internal to the external openings across the distal internal sphincter, but may extend proximally in the intersphincteric plane to end blindly with or without an abscess, or enter the rectum at a second internal opening. The fistulous track lies in the intersphincteric plane. The external opening is near the anal verge. These fistulas are treated by simple distal internal sphincterotomy similar to the lateral internal sphincterotomy done for anal fissure, and this procedure results in minor disturbance of continence. Ligation of the intersphincteric fistula tract is other method used to treat this type of fistula and recurrent anal fistulas (Xu and Tang, 2017). In addition, fistulas may be managed with laser ablation plus definitive flap closure of the internal fistula opening using the radial fiber Fistula laser Closing (FiLaC) device (Wilhelm et al., 2017) (Figs. 8 and 9).

3.4.2 Transsphincteric Fistula

Transsphincteric fistulas constitute about 40% of all fistulas that have a primary track that crosses both internal and external sphincters and which then

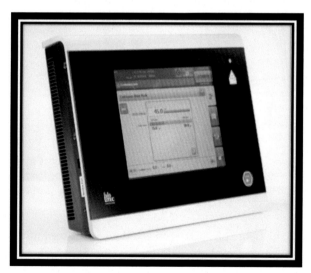

Fig. 8 Leonardo DUAL 45 diode laser from Biolitec AG, Germany. Wavelength 980–1470 nm. Maximum energy 15 W. *(From Wilhelm, A., Fiebig, A., Krawczak, M., 2017. Five years of experience with the FiLaC™ laser for fistula-in-ano management: long-term follow-up from a single institution. Tech. Coloproctol 21, 269–276.)*

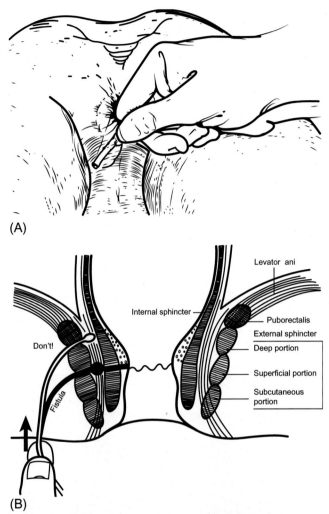

(A)

(B)

Fig. 9 (A) Palpation of fistula tract. (B) Probing the fistula tract. *(From Madoff, R.D., Goldberg, S.M., 1992. Reoperation for recurrent anal and perianal problems. In: McQuarrie, D.G., Humphrey, E.W. (Eds.), Preoperative General Surgery. Mosby, pp. 442–444 (Chapter 20).)*

passes through the ischioanal fossa to reach the skin of the buttock. The primary track may have secondary tracks arising from it, which often reach the roof of the ischioanal fossa, which may rarely pass through the levators to reach the pelvis, and which may spread circumferentially. In addition, the fistulous track starts in the intersphincteric plane or deep postanal space and traverses the external sphincter to open externally at the ischioanal fossa.

It includes the horseshoe fistula. The optimal surgical treatment for high transsphincteric fistulae-in-ano should attain complete eradication of the fistulous track and, at the same time, avoid compromise of the anal sphincters. Treatment is by placement of draining seton for high cryptoglandular anal fistula. Significant risk factors for recurrence of this fistula are previous fistula surgery, anterior anal fistula, and the presence of secondary tracks or branches as supralevator extension, and horseshoe fistula (Emile et al., 2017).

3.4.3 Suprasphincteric Fistula

Suprasphincteric fistulae are very rare and may be iatrogenic and are difficult to distinguish from high-level transsphincteric fistula. The fistulous track starts in the intersphincteric space at the dentate line then passes upward to reach above the puborectalis, then passes between the puborectalis and levator ani to go downward and open at the ischioanal fossa. Transsphincteric and suprasphincteric fistulas are difficult to treat. In general, treatment may or may not employ sphincterotomy.

3.4.4 Extrasphincteric Fistula

This type of fistula runs without specific relation to the sphincters and usually results from pelvic disease or trauma. The fistulous track starts at the rectal wall and traverses the levator ani and ischioanal fossa to open in the perianal skin. Its origin may be cryptogenic or traumatic, or due to intraabdominal pathology like diverticular disease and appendiceal abscess. Treatment of this fistula is a complicated procedure, especially closure of the primary opening of the fistula. Supralevator extrasphincteric fistulae-in-ano are rare as supralevator extension almost always is in the intersphincteric plane (Garg, 2017a, b, c).

The relationship of the external opening to the internal opening is suggested by Goodsall's rule. When the external opening is anterior to a line horizontally across the anus, then the track is most probably a simple track joining the internal opening, while if the external opening is posterior to this line, it is most probably a curvilinear and complex one (Fig. 10).

Treatment of a fistula is surgical in all cases, and is more complex due to the possibility of fecal incontinence as a result of sphincterotomy. Primary fistulotomy and cutting setons have the same incidence of fecal incontinence depending on the complexity of the fistula. So, the aim of a surgical procedure is to cure a fistula, and engaging in conservative management short of a major sphincterotomy is warranted to preserve fecal continence. However,

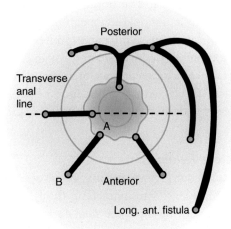

Fig. 10 Goodsall's rule. *(From Nelson, H., Cima, R.R., 2008. Anus. In: Townsend, C.M., Beauchamp, R.D., Evers, B.M., Mattox, K.L. (Eds.), Sabiston Text Book of Surgery, 18th ed. Saunders-Elsevier, pp. 1448 (Chapter 51).)*

trading radical surgery for conservative (nonsphincter-cutting) procedures such as a draining seton, fibrin sealant, anal fistula plug, endorectal advancement flap, dermal island flap, anoplasty, and LIFT (ligation of intersphincteric fistula tract) procedures all result in more recurrence/persistence, requiring repeated operations in many cases. A surgeon dealing with fistulas on a regular basis must tailor various operations to the needs of the patient depending upon the complexity of the fistula encountered (Abcarian, 2011).

3.5 Categorization of Fistulae-in-Ano

Fistulae-in-ano are classified as to grade according to increasing complexity, which can help guide their management. The classifications used are Parks, St. James Hospital University (SJHU), and Standard Practice Task Force (SPTF). Laying open (fistulotomy) of the fistula tract is the most common procedure for fistula-in-ano, and it has a high success rate. Thus, the description of a fistula as high, indicating a high risk of incontinence if laid open, or low, with a lower but still significant risk to function, is often used. So, the lower grade fistulas are supposed to have low risk of incontinence when laid open, and vice versa (Garg, 2017a, b, c). Similarly, fistulae-in-ano could be called simple or complex, with the term complex being a variation of the Parks classification, which describes fistulas whose management poses an increased risk for impairment of continence. Complexity may be endowed by the level at which the primary track crosses the sphincters, the presence of secondary extensions, or the difficulties faced in treatment. An anal fistula may be termed "complex" when the track crosses >30%–50% of the external sphincter (high-transsphincteric, suprasphincteric, and extrasphincteric), is anterior in a female, has multiple tracks, is recurrent, or the patient has preexisting incontinence, local irradiation, or Crohn's disease (Parks and Stitz, 1976). A new classification is being proposed which divides fistulae-in-ano into 5 grades in order of increasing complexity. Grade I and II are simple fistulas (fistulotomy may be done conveniently) and Grade III–V are high, complex fistulas (fistulotomy should not be attempted) (Garg, 2017a, b, c).

3.6 Treatment of Simple Fistulae-in-Ano

Fistulae-in-ano can be treated by fistulotomy or fistulectomy (Kronborg, 1985). Fistulotomy involves laying open all the fistulous track (unroofing) from the internal opening to the skin openings to establish drainage (Jain et al., 2012) (Fig. 11).

To start, we must identify the external and the internal openings and the track in-between (Uraiqat et al., 2010). The external opening is easy and obvious in the skin, while the internal opening is sometimes difficult to identify. Injection of hydrogen peroxide or dilute methylene blue may be

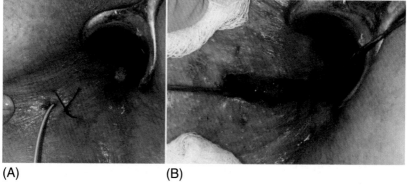

(A)   (B)

Fig. 11 Demonstrating technique of laying open fistula. (A) Insertion of probe and incision of tissue overlying probe. (B) Curettage of granulation tissue. *(From Nelson, H., Cima, R.R., 2008. Anus. In: Townsend, C.M., Beauchamp, R.D., Evers, B.M., Mattox, K.L. (Eds.), Sabiston Text Book of Surgery, 18th ed. Saunders-Elsevier, pp. 1449 (Chapter 51).)*

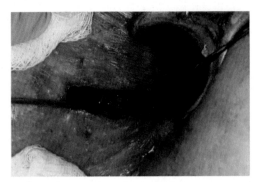

Fig. 12 Fisulotomy where the tract is probed and being transacted. *(From Nelson, H., Cima, R.R., 2008. Anus. In: Townsend, C.M., Beauchamp, R.D., Evers, B.M., Mattox, K.L. (Eds.), Sabiston Text Book of Surgery, 18th ed. Saunders-Elsevier, pp. 1449 (Chapter 51).)*

helpful. A probe is inserted in the fistulous track but care must be taken not to create a false track, and the track is opened over the probe entirely. Failing to do so may lead to recurrence. The granulation tissue filling the track is curetted and the edges marsupialized. Some surgeons practice excision of the whole track (fistulectomy), but this procedure has no great advantage over the fistulotomy procedure (Goodsall, 1900) (Fig. 12).

Horseshoe fistulas usually start in the deep postanal space and are an uncommon form of fistula-in-ano. If we attempt unroofing, we may end with long wound which is found to be unnecessary, so a deep incision in the postanal space with curettage of the lateral tracks is considered sufficient.

Laying open the fistulous track may involve transection of internal and external sphincter muscles. If the fistula is in the anterior half, transection of the sphincters may affect continence in older people (especially in women), but it is less harmful to continence in young patients (Vasilevsky and Gordon, 2007).

Simple anal fistulas were treated with track debridement and fibrin glue injection with few side effects and little risk of fecal incontinence (Abel et al., 1993). Fistulotomy yielded better results than fistulectomy regarding decreased duration of wound healing and duration of surgery without increasing the incidence of recurrence, incontinence, and postoperative pain (Murtaza et al., 2017).

3.7 Treatment of Complex Fistulae-in-Ano

The treatment of complex fistulas remains a major therapeutic challenge, balancing the risk of incontinence against the chance of permanent closure.

The treatment can be achieved by debridement and fibrin glue injection, with an endorectal advancement flap closure that can obliterate the septic focus and close the internal opening, does not divide the sphincter, results in a smaller wound, and can be combined with overlapping sphincter reconstruction for anterior fistulas.

Fistulas can be treated by curettage of the track, and fibrin glue filling of the track is done with closure of the internal opening. Fibrin glue has few side effects, but its overall success rate is <50%, especially in complex fistulas (Cintron et al., 2000).

Regarding the endorectal advancement flap, another surgical option is to close the internal fistula opening with an advancement flap. Advancement flap repairs result in high success rates with minimal effects on continence (Sonoda et al., 2002) (Fig. 13).

Other method is using a seton and/or staged fistulotomy. A seton is a silastic vessel loop made of flexible foreign-body-like suture material that is placed through the fistula track and secured to itself. Setons induce perisphincteric fibrosis along the fistula track, so that when the fistulotomy is eventually performed, or the seton gradually tightened, the muscular defect and amount of incontinence is limited (Pearl et al., 1993).

This is used when the fistulous opening is higher than the anorectal ring, and laying open the track may involve transection of the sphincters with an inevitable incontinence. The procedure should be staged with use of seton thread drawn though the fistula and tied loosely across the muscles, covering the fistula to allow drainage of infection and/or induce fibrosis. After 6–8 weeks, the seton is removed, and fistulotomy with curettage of epithelialized tissue is done with less risk of incontinence because scarring prevents retraction of the sphincter. A seton may be utilized to facilitate staged fistulotomy, to mark the external sphincter for later division after the subcutaneous components have healed (Williams et al., 1991; Garcia-Aguilar et al., 1998).

Another method of treatment for fistulas is using the staged drainage seton method combined with internal opening closure and relocation of the seton with 3–0 nylon, then, after a period of time, removing the 3–0 nylon. This method leads to a reduced rate of recurrence and incontinence (Figs. 14 and 15) (Lim et al., 2012).

Patients with high transsphincteric and suprasphincteric anocutaneous fistulas may be treated with a nitinol proctology clip for closure of this complex anocutaneous fistula. Healing rates for this method were comparable with those of other noninvasive, sphincter-sparing techniques for high,

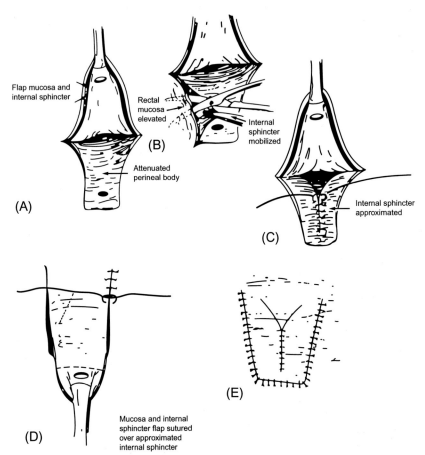

Fig. 13 Procedure of anorectal advancement flap. (A) Transsphincteric fistula. (B) Enlargement of external opening, curettage of tract, and outline of rectal flap. (C) Suture of internal opening, reflected rectal flap, and excision of flap apex. (D) Advancement of flap beyond internal opening and (E) suturing to anal canal. *(From Madoff, R.D., Goldberg, S.M., 1992. Reoperation for recurrent anal and perianal problems. In: McQuarrie, D.G., Humphrey, E.W. (Eds.), Preoperative General Surgery. Mosby, pp. 442–444 (Chapter 20).)*

complex, anocutaneous fistulas, with no risk of incontinence (Nordholm-Carstensen et al., 2017).

In India, patients were treated by ksharasutra (parasurgical procedure using a thread treated by alkalides) particularly for bhagandara, (fistula in ano) that was difficult to treat. Ksharasutra was prepared as per the ayurvedic pharmacopeia of India and used to treat the different cases of bhagandara. During treatment, panchawalkala kwatha (a decoction of five

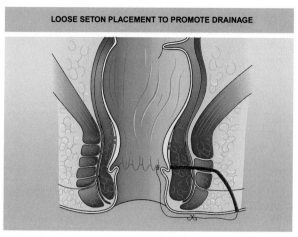

Fig. 14 Seton in place. *(From Fazio, V.W., Maher, J.W., Williamson, R.C.N., 2001. Gastrointestinal surgery. In: Corson, J.D., Williamson, R.C.N. (Eds.), Surgery. Mosby, pp. 21:13 (Section 3, Chapter 21).)*

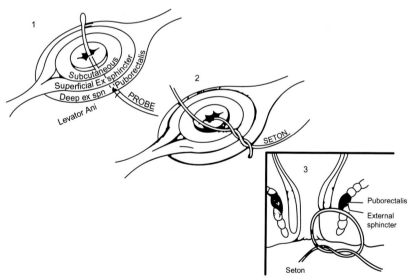

Fig. 15 This demonstrates the seton through the transsphincteric fistulous tract, and, on the left, shows completion of the rerouting procedure with repair of the incised internal anal sphincter and closure of the internal opening. *(From Madoff, R.D., Goldberg, S.M., 1992. Reoperation for recurrent anal and perianal problems. In: McQuarrie, D.G., Humphrey, E.W. (Eds.), Preoperative General Surgery. Mosby, pp. 442–444 (Chapter 20).)*

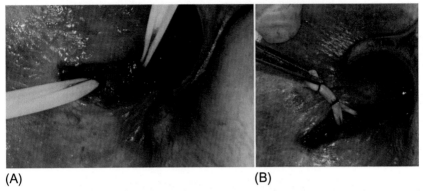

(A)   (B)

Fig. 16 Instead of using seton as in this patient with fistula, some times will be treated with ksharasutra threading. *(From Nelson, H., Cima, R.R., 2008. Anus. In: Townsend, C.M., Beauchamp, R.D., Evers, B.M., Mattox, K.L. (Eds.), Sabiston Text Book of Surgery, 18th ed. Saunders-Elsevier, pp. 1449 (Chapter 51).)*

medicinal plants' bark), shatdhautaghrita, jatyaditaila, and erandabhrishta-haritaki churna (powder) were used as adjuvant drugs. The patients were free of fistula, and no recurrence was observed. This treatment was reported to be safe and well tolerated in all the patients Fig. 16 (Nema et al., 2017).

4 ANAL FISTULA ASSOCIATED WITH CROHN'S DISEASE

Resembling the classical type of fistula, these may be simple and superficial or complex with multiple tracks, and high in origin from the lower rectum or upper anal canal. They manifest in the form of an indurated external opening exuding pus and a palpable track passing towards the anal canal.

These differ in treatment from ordinary fistula in that medical treatment should precede surgery. Drug treatment may include metronidazol and ciprofloxacin, which have been found to be useful in some patients. Some find methotrexate and azathioprine of benefit in treatment. Infliximab, which is an antibody directed against the tumor necrosis factor, has shown promising efficacy for fistulas associated with Crohn's disease (Matsuzawa et al., 2016). Surgical treatment, which should take into consideration the poor healing common among these patients, may involve local curetting of the granulation tissue, laying open superficial tracks, cleaning abscess cavities, and the use of an indwelling seton in higher fistulae as a method of surgical treatment (Faucheron et al., 1996; Sangwan et al., 1996; Michelassi et al., 2000).

REFERENCES

Abcarian, H., 2011. Anorectal infection: abscess–fistula. Clin. Colon Rectal Surg. 24, 14–21.

Abel, M.E., Chiu, Y.S., Russell, T.R., Volpe, P.A., 1993. Autologous fibrin glue in the treatment of rectovaginal and complex fistulas. Dis. Colon Rectum 36, 447–449.

Acharya, J.T., 1994a. Sushruta Samhita, reprint. Chikitsa Sthana, 17/29-32. In: Sushruta (Ed.), Varanasi: Chowkhambha Surabharati Prakashan, p. 378.

Acharya, J.T., 1994b. Sushruta Samhita, reprint. Sutra Sthana, 11/4. In: Sushruta (Ed.), Varanasi: Chowkhambha Surabharati Prakashan, p. 40.

Belliveau, P., 1990. Anal fistula. Current Therapy in Colon and Rectal Surgery. BC Decker, Philadelphia, pp. 22–27.

Bubrick, M., Hitchcock, C., 1979. Necrotizing anorectal and perianal infections. Surgery 86, 655–662.

Calata, C.L., Teoule, P., Bussen, D.G., 2017. Perianal pain and swelling. Chirurg 21.

Chrabot, C.M., Prasad, M.L., Abcarian, H., 1983. Recurrent anorectal abscesses. Dis. Colon Rectum 26, 105–108.

Cintron, J.R., Park, J.J., Orsay, C.P., et al., 2000. Repair of fistulasin- ano using fibrin adhesive: long-term follow-up. Dis. Colon Rectum 43, 944–950.

Corman, M.L., 1993. Anorectal Abscess and Anal Fistula. Colon and Rectal Surgery. JB Lippincott, Philadelphia, pp. 133–187.

Corman, M.L., 2005. Anal Fistula. Colon & Rectal Surgery, fifth ed. Lippincott Williams & Wilkins, Philadelphia, PA (Chapter 11).

Dwivedy, R., 2011. Chakradatta, reprint. Arsha Chikitsa 5/148. In: Chakrapanidatta (Ed.), Varanasi: Chowkhambha Sanskrit Bhavan, p. 66.

Emile, S.H., Elfeki, H., Thabet, W., Sakr, A., Magdy, A., El-Hamed, T.M.A., Omar, W., Khafagy, W., 2017. Predictive factors for recurrence of high transsphincteric anal fistula after placement of seton. J. Surg. Res. 1, 261–268.

Faucheron, J.L., Saint-Marc, O., Guibert, L., Parc, R., 1996. Longterm seton drainage for high anal fistulas in Crohn's disease–a sphincter-saving operation? Dis. Colon Rectum 39, 208–211.

Faujdar, H.S., Mehta, G., Agarwal, R.K., Malpani, N.K., 1981. Management of fistula in ano. J. Postgrad. Med. 27, 172b–177.

Fazio, V.W., Maher, J.W., Williamson, R.C.N., 2001. Gastrointestinal surgery. In: Corson, J.D., Williamson, R.C.N. (Eds.), Surgery. Mosby, pp. 21:13 (Section 3, Chapter 21).

Garcia-Aguilar, J., Belmonte, C., Wong, D.W., Goldberg, S.M., Madoff, R.D., 1998. Cutting seton versus two-stage seton fistulotomy in the surgical management of high anal fistula. Br. J. Surg. 85, 243–245.

García-Granero, A., Granero-Castro, P., Frasson, M., Flor-Lorente, B., Carreño, O., Garcia-Granero, E., 2014. The use of an endostapler in the treatment of supralevator abscess of intersphincteric origin. Color. Dis. 16, O335–8.

Garcia-Granero, A., Granero-Castro, P., Frasson, M., Flor-Lorente, B., Carreño, O., Espí, A., Puchades, I., Garcia-Granero, E., 2014. Management of cryptoglandular supralevator abscesses in the magnetic resonance imaging era: a case series. Int. J. Color. Dis. 29, 1557–1564.

Garg, P., 2017a. Transanal opening of intersphincteric space (TROPIS)—a new procedure to treat high complex anal fistula. Int. J. Surg. 40, 130–134.

Garg, P., 2017b. Supralevator extrasphincteric fistula-in-ano are rare as supralevator extension is almost always in the intersphincteric plane. World J. Surg. (Epub ahead of print).

Garg, P., 2017c. Comparing existing classifications of fistula-in-ano in 440 operated patients: is it time for a new classification? A retrospective cohort study. Int. J. Surg. 42, 34–40.

Goldberg, S., Gordon, P., Nivatvongs, S., 1980a. Essentials of Anorectal Surgery. JB Lippin-
   cott, Philadelphia, p. 103.
Goldberg, S.M., Gordon, P.H., Nivatvongs, S., 1980b. Essentials of Anorectal Surgery. JB
   Lippincott, Philadelphia, pp. 100–127.
Goodsall, D.M., 1900. Anorectal fistula. In: Goodsall, D.M., Miles, W.E. (Eds.), Diseases of
   Anus and Rectum. 92. Longman and Green, London.
Gordon, P., 1992. Anorectal abscess and fistula in ano. In: Gordon, P., Nivatvongs, S. (Eds.),
   Principles and Practice of Surgery of the Colon, Rectum, and Anus. Quality Medical
   Publishing, St. Louis, p. 221.
Grace, R.H., 1990. The management of acute anorectal sepsis. Ann. R. Coll. Surg. Engl.
   72160.
Hanley, P.H., Ray, J.E., Pennington, E.E., Grablowsky, O.M., 1976. A ten year follow up
   study of horseshoe-abscess fistula-in-ano. Dis. Colon Rectum 19, 507–515.
Hasan, R.M., 2016. A study assessing postoperative corrugate rubber drain of perianal
   abscess. Ann. Med. Surg. 11, 42–46.
Hasan, R.M., 2017. Postoperative empirical antibiotic use for uncomplicated perianal abscess
   and fistula. Ann. Colorectal Res. 5, e40795.
Hopping, R.A., 1956. Management of ischioanal abscess and fistula in ano. J. Med. Soc. N. J.
   53, 135–137.
Ibn Majdoub Hassani, K., Ait Laalim, S., Toughrai, I., Mazaz, K., 2012. Perianal Tubercu-
   losis: A Case Report and a Review of the Literature. Hindawi Publishing Corporation,
   Morocco, p. 4.
Jain, B.K., Vaibhaw, K., Garg, P.K., Gupta, S., Mohanty, D., 2012. Comparison of a fistu-
   lectomy and a fistulotomy with marsupialization in the management of a simple anal fis-
   tula: a randomized, controlled pilot trial. J. Korean Soc. Coloproctol. 28, 78–82.
Kodner, I.J., Mazor, A., Shemesh, E.I., Fry, R.D., Fleshman, J.W., Birnbaum, E.H., 1993.
   Endorectal advancement flap repair of rectovaginal and other complicated anorectal fis-
   tulas. Surgery 114, 682–690.
Kronborg, O., 1985. To lay open or excise a fistula-in-ano: a randomized trial. Br. J. Surg.
   72, 970.
Kronborg, O., Olsen, H., 1984. Incision and drainage v. incision, curettage and suture under
   antibiotic cover in anorectal abscess. A randomized study with 3-year follow-up. Acta
   Chir. Scand. 150, 689–692.
Kurapati, V.K., Nishteswar, K., 2014. Management of ano–rectal disorders by *Kṣārasūtra*: a
   clinical report. Anc. Sci. Life 34, 89–95.
Lim, C.H., Shin, H.K., Kang, W.H., Park, C.H., Hong, S.M., Jeong, S.K., Kim, J.Y.,
   Yang, H.K., 2012. The use of a staged drainage seton for the treatment of anal fistulae
   or fistulous abscesses. J. Korean Soc. Coloproctol. 28, 309–314.
Lohsiriwat, V., 2016. Anorectal emergencies. World J. Gastroenterol. 14, 5867–5878.
Lohsiriwat, V., Yodying, H., Lohsiriwat, D., 2010. Incidence and factors influencing the
   development of fistula-in-ano after incision and drainage of perianal abscesses. J. Med.
   Assoc. Thail. 93, 61–65.
Madoff, R.D., Goldberg, S.M., 1992. Reoperation for recurrent anal and perianal problems.
   In: McQuarrie, D.G., Humphrey, E.W. (Eds.), Preoperative General Surgery. Mosby,
   pp. 442–444, 448 (Chapter 20).
Malik, A.I., Nelson, R.L., Tou, S., 2010. Incision and drainage of perianal abscess with or
   without treatment of anal fistula. Cochrane Database Syst. Rev. 7, CD006827.
Matsuzawa, F., Homma, S., Yoshida, T., Shibasaki, S., Minagawa, N., Shimokuni, T.,
   Sakihama, H., Kawamura, H., Takahashi, N., Taketomi, A., 2016. Successful treat-
   ment of rectovaginal fistula and rectal stenosis due to perianal Crohn's disease by
   dual-port laparoscopic abdominoperineal resection: a report of two cases. Surg. Case
   Rep. 2, 83.

Michelassi, F., Melis, M., Rubin, M., Hurst, R.D., 2000. Surgical treatment of anorectal complications in Crohn's disease. Surgery 128, 597–603.

Millan, M., García-Granero, E., Esclápez, P., Flor-Lorente, B., Espí, A., Lledó, S., 2006. Management of intersphincteric abscesses. Color. Dis. 8, 777–780.

Murtaza, G., Shaikh, F.A., Chawla, T., Rajput, B.U., Shahzad, N., Ansari, S., 2017. Fistulotomy versus fistulectomy for simple fistula in ano: a retrospective cohort study. J. Pak. Med. Assoc. 67, 339–342.

Nelson, H., Cima, R.R., 2008. Anus. In: Townsend, C.M., Beauchamp, R.D., Evers, B.M., Mattox, K.L. (Eds.), Sabiston Text Book of Surgery, 18th ed. Saunders-Elsevier, pp. 1446–1449 (Chapter 51).

Nelson, R.L., Prasad, M.L., Abcarian, H., 1985. Anal carcinoma presenting as a perirectal abscess or fistula. Arch. Surg. 120, 632–635.

Nema, A., Gupta, S.K., Dudhamal, T.S., Mahanta, V., 2017. Transrectal ultra sonography based evidence of Ksharasutra therapy for Bhagandara (Fistula-in-ano)—a case series. J. Ayurveda Integr. Med. 6. pii: S0975-9476-30159.

Nordholm-Carstensen, A., Krarup, P.M., Hagen, K., 2017. Treatment of complex fistula-in-ano with a nitinol proctology clip. Dis. Colon Rectum 60, 723–728.

O'Malley, G.F., Dominici, P., Giraldo, P., Aguilera, E., Verma, M., Lares, C., Burger, P., Williams, E., 2009. Routine packing of simple cutaneous abscesses is painful and probably unnecessary. Acad. Emerg. Med. 16, 470–473.

Ommer, A., Herold, A., Berg, E., Fürst, A., Post, S., Ruppert, R., Schiedeck, T., Schwandner, O., Strittmatter, B., 2017. German S3 guidelines: anal abscess and fistula (second revised version). Langenbeck's Arch. Surg. 402, 191–201.

Ortega, A.E., Bubbers, E., Liu, W., Cologne, K.G., Ault, G.T., 2015. A novel classification, evaluation, and treatment strategy for supralevator abscesses. Dis. Colon Rectum 58, 1109–1110.

Parks, A.G., Stitz, R.W., 1976. The treatment of high fistula-in-ano. Dis. Colon Rectum 19, 487–499.

Parks, A.G., Gordon, P.H., Hardcastle, J.D.A., 1976. Classification of fistula-in-ano. Br. J. Surg. 63, 1–12.

Pearl, R.K., Andrews, J.R., Orsay, C.P., et al., 1993. Role of the seton in the management of anorectal fistulas. Dis. Colon Rectum 36, 573–579.

Perera, A.P., Howell, A.M., Sodergren, M.H., Farne, H., Darzi, A., Purkayastha, S., Paraskeva, P., 2015. A pilot randomised controlled trial evaluating postoperative packing of the perianal abscess. Langenbeck's Arch. Surg. 400, 267–271.

Read, D.R., Abcarian, H., 1979. A prospective survey of 474 patients with anorectal abscess. Dis. Colon Rectum 22, 566–568.

Sangwan, Y.P., Schoetz Jr., D.J., Murray, J.J., Roberts, P.L., Coller, J.A., 1996. Perianal Crohn's disease: results of local surgical treatment. Dis. Colon Rectum 39, 529–535.

Schwartz, D.A., Wiersema, M.J., Dudiak, K.M., et al., 2001. A comparison of endoscopic ultrasound, magnetic resonance imaging, and exam under anesthesia for evaluation of Crohn's perianal fistulas. Gastroenterology 121, 1064–1072.

Sonoda, T., Hull, T., Piedmonte, M.R., Fazio, V.W., 2002. Outcomes of primary repair of anorectal and rectovaginal fistulas using the endorectal advancement flap. Dis. Colon Rectum 45, 1622–1628.

Stewart, M.P., Laing, M.R., Krukowski, Z.H., 1985. Treatment of acute abscesses by incision, curettage and primary suture without antibiotics: a controlled clinical trial. Br. J. Surg. 72, 66–67.

Tonkin, D.M., Murphy, E., Brooke-Smith, M., Hollington, P., Rieger, N., Hockley, S., Richardson, N., Wattchow, D.A., 2004. Perianal abscess: a pilot study comparing packing with nonpacking of the abscess cavity. Dis. Colon Rectum 47, 1510–1514.

Uraiqat, A., Al-Shobaki, M., Al-Rashaidah, M., 2010. Fistula-in-Ano: a prospective audit. JRMS 17, 43–49.

Vasilevsky, C.A., Gordon, P.H., 2007. Benign anorectal sepsis. In: The ASCRS Text Book of Colon and Rectal Surgery. Vol. 13. 192 pp.

Ward, T., Snooks, S.J., Croft, R.J., 1987. Perianal pain and swelling due to a pre-coccygeal tumor. J. R. Soc. Med. 80, 651–652.

Whiteford, M.H., Kilkenny, J., Hyman, N., Buie, W.D., Cohen, J., Orsay, C., Dunn, G., Perry, W.B., Ellis, C.N., Rakinic, J., Gregorcyk, S., Shellito, P., Nelson, R., Tjandra, J.J., Newstead, G., The Standards Practice Task Force, the American Society of Colon and Rectal Surgeons, 2005. Practice parameters for the treatment of perianal abscess and fistula-in-ano. Dis. Colon Rectum 48, 1337–1342.

Wilhelm, A., Fiebig, A., Krawczak, M., 2017. Five years of experience with the FiLaC™ laser for fistula-in-ano management: long-term follow-up from a single institution. Tech. Coloproctol 21, 269–276.

Williams, J.G., MacLeod, C.A., Rothenberger, D.A., Goldberg, S.M., 1991. Seton treatment of high anal fistulae. Br. J. Surg. 78, 1159–1161.

Williams, J.G., Farrands, P.A., Williams, A.B., Taylor, B.A., Lunniss, P.J., Sagar, P.M., et al., 2007. The treatment of anal fistula: ACPGBI position statement. Color. Dis. 9, 18–50.

Xu, Y., Tang, W., 2017. Ligation of intersphincteric fistula tract is suitable for recurrent anal fistulas from follow-up of 16 months. Biomed. Res. Int. 2017, 3152424.

FURTHER READING

Beck, D., Wexner, S.D., 1992. Fundamentals of Anorectal Surgery. McGraw-Hill, Inc., New York.

World Health Organization. (n.d.) International Clinical Trial Registry Platform: http://www.who.int/trialsearch/trial.aspx?trialid=NCT01864473.

CHAPTER 7

Role of Antimicrobial Agents in the Management of Perianal Abscess

Batool Mutar Mahdi

Contents

1 INTRODUCTION

Perianal abscesses and anal fistulas are the acute and chronic manifestations of infection in the perianal area (Rizzo et al., 2010). They are characterized by the collection of pus to form an abscess in the perianal, intersphincteric, ischiorectal, or pelvirectal spaces (James Garden et al., 2002; Whiteford, 2007). These conditions often present with perianal swelling accompanied by severe pain and/or constipation, and may cause systemic infection and life-threatening sepsis (Goldberg et al., 1980). Most perianal abscesses are formed by occlusion of an anal gland duct with subsequent bacterial overgrowth and abscess formation (Parks, 1961). Principal management is through incision and adequate surgical drainage (ID) (Malik et al., 2010) Various strains of aerobic/anaerobic bacteria might be responsible for perianal abscess formation (Brook and Martin, 1980), with enteric bacteria being especially more common in perianal diseases (Niyogi et al., 2010). Abscesses and fistulas are usually treated with a proper selection of antibiotics and surgical drainage of the abscess cavity. Pus must be sent for culture and sensitivity, but most surgeons tend not to review swab culture, and sensitivity results (Leung et al., 2009).

2 OBJECTIVES OF USING ANTIBIOTICS IN MANAGEMENT

The use of broad-spectrum antibiotics for perianal disease before or after management remains common, although with questionable benefit (Stewart et al., 1985). Previously, it was thought that the routine use of antibiotics following incision and drainage of an abscess had no impact on healing time or reduction of recurrence rates, and was, therefore, not ordinarily indicated (Macfie and Harvey, 1977). The current guidelines do not include an agreement on the use of antibiotics, and their routine use is not recommended; however, they are recommended for use in the presence of immunosuppressive conditions (Ommer et al., 2012). Regarding the management of perianal abscesses, the hallmark of good management is prompt surgery; however, antibiotics still have a role to play in specific cases. Antibiotics have only a small part to play, and they are unlikely to abort the infection once the symptoms have been present for 24 h, as they cannot penetrate into the pus, and there is often some necrosis of fatty tissue (Parchment Smith and Hernon, 2001). Thus, the main method of management is incision and drainage (Llera and Levy, 1985), and the practitioner should provide appropriate empiric intravenous antibiotic coverage for patients who are elderly or immunosuppressed, patients who have comorbidities, patients with a heart valve abnormality or prosthetic valve, previous bacterial endocarditis, complex congenital heart disease, surgically constructed systemic pulmonary shunts or conduits, congenital cardiac malformations, acquired valvular dysfunction (e.g., rheumatic heart disease), hypertrophic cardiomyopathy, and mitral valve prolapse with valvular regurgitation and/or thickened leaflets, who are likely to benefit from antibiotic prophylaxis, and patients in whom infection has become systemic (Dajani et al., 1997; Nunoo-Mensah et al., 2006). In addition, antibiotics should be used when abscess is caused by bowel flora organisms, whether in pure growth or a mixture of like bacteria belonging to the Enterobacteriacea family and/or anaerobes (Ulug et al., 2010). On the other hand, in certain situations, the infectious agent may be a Group A beta-hemolytic streptococcus or *Staphylococcus aureus*, which are more virulent than those of the bowel flora and have a greater propensity to invade and/or relapse, and hence, a worse infection outcome. In fact, some of these infections may be due to a multiresistant organism such as

MRSA or an even a more virulent strain of bacteria such as PVL-producing *S. aureus* (Park et al., 2008; Brown et al., 2009). It is, therefore, recommended that, in addition to prompt surgery, an appropriate antibiotic is prescribed as guided by the culture and sensitivity results or empirical use of antibiotics, and about 98% of positive swab cultures have been sensitive to routine empirical antibiotics (Seow-En and Ngu, 2014). In the interest of antimicrobial stewardship, a monospectrum antimicrobial should be used, as opposed to a broad-spectrum one. Hence, appropriate sampling of the abscess for microbiological investigations is advisable. In addition, antibiotics administered after incision and drainage reduced the rate of fistula formation, abscess recurrence, cellulitis, and sepsis. Empiric antibiotics have an important role in decreasing fistula formation and preventing recurrence (Hasan, 2017). The use of antibiotics reduces the development of fistula-in-ano after abscess drainage (Lohsiriwat et al., 2010; Afsarlar et al., 2011), while in other cases, it did not affect the results (Sözener et al., 2011). Regarding fistula-in-ano, a number of risks are likely to occur, such as infection which requires antibiotic administration (NHS Choices, n.d.). Studies done in the 1980s demonstrated that cultures and sensitivity for the draining pus was essential because certain gut-specific bacteria were indicative of the presence of a fistula-in-ano (Eykyn and Grace, 1986). This was specific in only 80% of the cases (Lunniss and Phillips, 1994). The use of preoperative antibiotics was not associated with development of fistula-in-ano or recurrence of abscess (Hamadani et al., 2009; Idris et al., 2011).

3 TYPES OF ANTIBIOTICS USED

Appropriate intravenous antibiotic coverage should be provided preoperatively and postoperatively, either on the basis of Gram staining, culture, and sensitivity, or on an empiric basis as a preventive measure, for elderly patients, patients with immunosuppression, patients with a heart valve abnormality or prosthesis, diabetic patients, and patients with comorbid states (Steele et al., 2011; Wilson et al., 2007).

3.1 Amoxycillin/Clavulanic Acid

Amoxycillin/clavulanic acid is a combination of two drugs which interferes with bacterial cell-wall synthesis during active replication, causing bactericidal activity against susceptible organisms. Amoxycillin/clavulanic acid covers skin, enteric flora, and anaerobes; it is not ideal for nosocomial pathogens. Patients would have responded to common first-line antibiotics like amoxycillin/clavulanic acid and metronidazole (Seow-En and Ngu, 2014). The most common aerobic bacteria were *Escherichia coli*, and the most common anaerobic bacteria were *Bacteroides fragilis*. *E. coli* were sensitive to amoxicillin-clavulamic acid in (84.6%) of the cases (Cheng and Tsai, 2010).

3.2 Imipenem

Imipenem should be used empirically for more severely ill intensive care unit patients. Pus or blood culture and sensitivity results, once available, should guide antibiotic selection. Predisposing and comorbid diseases may also guide empiric antibiotic selection (Jeong et al., 2015).

3.3 Ampicillin

Ampicillin, a broad-spectrum penicillin, interferes with bacterial cell-wall synthesis during active replication, causing bactericidal activity against susceptible organisms. It is used as an alternative to amoxicillin. Patients with prosthetic heart valves who are at risk for endocarditis should receive IV prophylactic antibiotics before any procedure. Ampicillin is preferred for this application, unless the patient is penicillin-allergic, in which case cefazolin or clindamycin is an appropriate choice. It has been shown that there is a 13% abscess recurrence rate after incision, curettage, and primary closure of a perianal abscess using the antibiotics ampicillin and metronidazole for 1 to 4 days (Lundhus and Gotttrup, 1993).

3.4 Cefazolin

Cefazolin is a first-generation semisynthetic cephalosporin that binds to one or more penicillin-binding proteins, arresting bacterial cell-wall synthesis and inhibiting bacterial replication. It has poor capacity to cross the blood-brain barrier. Cefazolin is primarily active against skin flora, including *S. aureus*, and is typically used alone for skin and skin-structure coverage. Regimens for IV and intramuscular (IM) dosing are similar. This agent is used in penicillin-allergic patients with prosthetic heart valves who are at risk for endocarditis. The most common aerobic bacteria were *E. coli*, and the most common anaerobic bacteria were *B. fragilis*. *E. coli* were sensitive to cefazolin in 84.6% of the patients (Cheng and Tsai, 2010). Prophylactic use of Cefazolin antibiotics to the patients before surgery showed no recurrence (Akkapulu et al., 2015).

3.5 Clindamycin

Clindamycin is a semisynthetic antibiotic produced by 7(*S*)-chloro-substitution of the 7(R)-hydroxyl group of the parent compound, lincomycin. It inhibits bacterial growth, possibly by blocking dissociation of peptidyl tRNA from ribosomes, causing RNA-dependent protein synthesis to arrest. Clindamycin is widely distributed in the body without penetrating the central nervous system. It is protein-bound and excreted by the liver and kidneys. This agent is used in penicillin-allergic patients with prosthetic heart valves at risk for endocarditis. Lincomycin used in patients with perianal abscess showed an abscess recurrence rate of 10% and 5% after 6 and 12 months, respectively, while fistula formation after 6 and 12 months was 25% and 5%, respectively (Hasan, 2017). So, lincomycin used in the treatment of abscess and fistula postoperatively as an empiric antibiotic resulted in less recurrence of abscess or fistula development.

3.6 Metronidazole and Ciprofloxacin

Antibiotics (metronidazole and ciprofloxacin) had merely a short-term benefit in the closure of fistulas (Solomon et al., 1993). The most common aerobic bacteria were *E. coli*, and the most common anaerobic bacteria were *B. fragilis*. *E. coli* were sensitive to ciprofloxacin in 69.2% of the cases. For anaerobic bacteria (*Bacteroides* species and *Clostridium perfringens*), the antibiotic sensitivity rates were determined to be 100% for metronidazole (Cheng and Tsai, 2010). Metronidazole and fluoroquinolones demonstrated improved perianal symptoms in over 90% of patients with fistulizing disease (McKee and Keenan, 1996).

Thus, empirical broad-spectrum antibiotics with anaerobic coverage should be considered in the treatment of perianal disease for polymicrobial infection (Chen et al., 2013). An appropriate selection of antibiotics is important for empirical use before attaining the results of bacterial culture. The first choice of antibiotics includes augmentin and cefazolin for aerobes, metronidazole for anaerobes, and amikacin for severe infection. Gentamycin is not recommended in severe infection because some *E. coli* isolates from perianal abscesses exhibited gentamycin resistance (Cheng and Tsai, 2010).

REFERENCES

Afsarlar, C.E., Karaman, A., Tanır, G., Karaman, I., Yılmaz, E., Erdog an, D., et al., 2011. Perianal abscess and fistula-in-ano in children: clinical characteristic, management and outcome. Pediatr. Surg. Int.

Akkapulu, N., Dere, O., Zaim, G., Soy, H.E.A., Ozmen, T., Doğrul, A.B., 2015. A retrospective analysis of 93 cases with anorectal abscess in a rural state hospital. Turkish J. Surg. (Ulusal Cer Derg) 31, 5–8.

Brook, I., Martin, W.J., 1980. Aerobic and anaerobic bacteriology of perirectal abscess in children. Pediatrics 66, 282–284.

Brown, S.R., Horton, J.D., Davis, K.G., 2009. Perirectal abscess infections related to MRSA: a prevalent and underrecognized pathogen. J. Surg. Educ. 66, 264–266.

Chen, C.-Y., Cheng, A., Huang, S.-Y., Sheng, W.-H., Liu, J.-H., Ko, B.-S., Yao, M., Chou, W.-C., Lin, H.-C., Chen, Y.-C., Tsay, W., Tang, J.-L., Chang, S.-C., Tien, H.-F., 2013. Clinical and microbiological characteristics of perianal infections in adult patients with acute leukemia. PLoS One 8, e60624.

Cheng, S.F., Tsai, W.S., 2010. Microbiological analysis of perianalabscess and its treatment. J. Soc. Colon Rectal Surgeon 21, 37–42.

Dajani, A.S., Taubert, K.A., Wilson, W., Bolger, A.F., Bayer, A., Ferrieri, P., et al., 1997. Prevention of bacterial endocarditis. Recommendations by the American Heart Association. Circulation 96, 358–366.

Eykyn, S.J., Grace, R.H., 1986. The relevance of microbiology in the management of anorectal sepsis. Ann. R. Coll. Surg. Engl. 68, 237–239.

Goldberg, S., Gordon, P., Nivatvongs, S., 1980. Essentials of Anorectal Surgery. JB Lippincott, Philadelphia, p. 103.

Hamadani, A., Haigh, P.I., Liu, I.L., et al., 2009. Who is at risk for developing chronic anal fistula or recurrent anal sepsis after initial perianal abscess? Dis. Colon Rectum 52, 217–221.

Hasan, R.M., 2017. Postoperative empirical antibiotic use for uncomplicated perianal abscess and fistula. Ann. Colorectal Res. 28, e40795.

Idris, S.A., Hamza, A.A., Alegail, I.M.A., 2011. The relation between the presence of intestinal bacteria in the perianal abscess and the anticipated perianal fistula. Sudan J. Med. Sci. 6, 199–208.

James Garden, O., Bradbury, A.W., Forsythe, J., 2002. Principles and Practice of Surgery, fourth ed., pp. 357–360.

Jeong, W.S., Choi, S.Y., Jeong, E.H., Bang, K.B., Park, S.S., Lee, D.S., Park, D., Jung, Y.S., 2015. Perianal abscess and proctitis by Klebsiella pneumonia. Intest. Res. 13, 85–89.

Leung, E., McArdle, K., Yazbek-Hanna, M., 2009. Pus swabs in incision and drainage of perianal abscesses: what is the point? World J. Surg. 33, 2448–2451.

Llera, J.L., Levy, R.C., 1985. Treatment of cutaneous abscess: a double-blind clinical study. Ann. Emerg. Med. 14, 15–19.

Lohsiriwat, V., Yodying, H., Incidence, L.D., 2010. Factors influencing the development of fistula-in-ano after incision and drainage of perianal abscesses. J. Med. Assoc. Thail. 93, 61–65.

Lundhus, E., Gotttrup, F., 1993. Outcome at three to five years of primary closure of perianal and pilonidal abscess. A randomized, double-blind clinical trial with a complete three-year followup of one compared with four days treatment with ampicillin and metronidazole. Eur. J. Surg. 159, 555–558.

Lunniss, P.J., Phillips, R.K., 1994. Surgical assessment of acute anorectal sepsis is a better predictor of fistula than microbiological analysis. Br. J. Surg. 81, 368–369.

Macfie, J., Harvey, J., 1977. The treatment of acute superficial abscesses: a prospective clinical trial. Br. J. Surg. 64, 264–266.

Malik, A.I., Nelson, R.L., Tou, S., 2010. Incision and drainage of perianal abscess with or without treatment of anal fistula. Cochrane Database Syst. Rev.(7), CD006827.

McKee, R.F., Keenan, R.A., 1996. Perianal Crohn's disease: is it all bad news? Dis. Colon Rectum 39, 136–142.

NHS Choices, Available from: www.nhs.uk/Conditions/Analfistula/Pages/Treatment. aspx.

Niyogi, A., Agarwal, T., Broadhurst, J., Abel, R.M., 2010. Management of perianal abscess and fistula-in-ano in children. Eur. J. Pediatr. Surg. 20, 35–39.

Nunoo-Mensah, J.W., Balasubramaniam, S., Wasserberg, N., Artinyan, A., Gonzalez-Ruiz, C., Kaiser, A.M., et al., 2006. Fistula-in-ano: do antibiotics make a difference? Int. J. Color. Dis. 21, 441–443.

Ommer, A., Herold, A., Berg, E., Furst, A., Sailer, M., Schiedeck, T., 2012. German S3 guideline: anal abscess. Int. J. Color. Dis. 27, 831–837.

Parchment Smith, C., Hernon, C., 2001. Anal and perianal conditions. In: MRCS System Modules: Essential Revision Notes, second ed, pp. 517–547.

Park, C., et al., 2008. A case of perianal abscess due to Panton-valentine Leukocidin positive community-associated methicillin-resistant Staphylococcus aureus. Infect. Chemother. 40, 121–126.

Parks, A.G., 1961. Pathogenesis and treatment of fistuila-in-ano. Br. Med. J. 1 (5224), 463–469.

Rizzo, J.A., Naig, A.L., Johnson, E.K., 2010. Anorectal abscess and fistula-in-ano: evidence-based management. Surg. Clin. North Am. 90, 45–68.

Seow-En, I., Ngu, J., 2014. Routine operative swab cultures and post-operative antibiotic use for uncomplicated perianal abscesses are unnecessary. ANZ J. Surg.

Solomon, M.J., McLeod, R.S., Connor, B.I., 1993. Combination ciprofloxacin and metronidazole in severe perianal Crohn's disease. Can. J. Gastroenterol. 7, 571–573.

Sözener, U., Gedik, E., Kessaf Aslar, A., Ergun, H., Halil Elhan, A., Memikoglu, O., et al., 2011. Does adjuvant antibiotic treatment after drainage of anorectal abscess prevent development of anal fistulas? A randomized, placebo-controlled, double-blind, multi-center study. Dis. Colon Rectum 54, 923–929.

Steele, S.R., Kumar, R., Feingold, D.L., Rafferty, J.L., Buie, W.D., Standards Practice Task Force of The American Society of Colon and Rectal Surgeons, 2011. Practice parameters for the management of perianal abscess and fistula-in-ano. Dis. Colon Rectum 54, 1465–1474.

Stewart, M.P., Laing, M.R., Krukowski, Z.H., 1985. Treatment of acute abscesses by incision, curettage and primary suture without antibiotics: a controlled clinical trial. Br. J. Surg. 72, 66–67.

Ulug, M., et al., 2010. The evaluation of bacteriology in perianal abscesses of 81 adult patients. Braz. J. Infect. Dis. 14, 225–229.

Whiteford, M.H., 2007. Perianal abscess/fistula disease. Clin. Colon Rectal Surg. 20, 102–109.

Wilson, W., Taubert, K.A., Gewitz, M., Lockhart, P.B., Baddour, L.M., Levison, M., et al., 2007. Prevention of infective endocarditis: Guidelines from the American Heart Association: A guideline from the American Heart Association rheumatic fever, endocarditis, and Kawasaki disease committee, council on cardiovascular disease in the young, and the council on clinical cardiology, council on cardiovascular surgery and anesthesia, and the quality of care and outcomes research interdisciplinary working group. Circulation 116, 1736–1754.

CHAPTER 8

Complications of Perianal Diseases

Riyadh Mohammad Hasan

Contents

1 COMPLICATIONS

Proper treatment by incision and drainage of anal abscess and treatment of fistula will heal both without recurrence and, consequently, no further symptoms or suffering will remain. Otherwise, the patient will develop one or more of the following complications.

1.1 Complications Due to Delayed Surgical Intervention

1.1.1 Rupture With Subsequent Recurrences or Fistula Formation

Abscesses may rupture spontaneously and fistulas persist because the patients did not seek advice or go to a clinic for management due to embarrassment, or because the patient refused surgical intervention.

1.1.2 Necrotizing Fasciitis

This is a life-threatening complication which is common in immune-compromised and elderly patients who are 50 years and older. Incidence is reported to be as high as 0.40 in 100,000. Males are more commonly affected, at a ratio of 3:1. Infection usually spreads upward along the skin, subcutaneous tissue, fascia, and (rarely) through muscles. There is a rapid spread of infection with tissue necrosis, and severe pain ending in death.

Patients are usually managed by excessive surgical debridement of necrotic tissues with hyperbaric oxygen and antibiotics (Abcarian, 1982).

New Concepts in the Management of Septic Perianal Conditions
https://doi.org/10.1016/B978-0-12-816111-1.00008-2

1.2 Postoperative Complications

1.2.1 Early Complications

1. Postoperative bleeding rarely occurs and can be treated with simple packing.
2. Inadequate drainage of the abscess, that is, missing pockets of the abscess.
3. Missed abscesses in the case of horseshoe abscess or submucus abscess.
4. Urinary retention, hemorrhage, fecal impaction, and thrombosed external hemorrhoids, which were found to occur in <6% of cases (Mazier, 1971).

1.2.2 Late Complications

Pain, bleeding, pruritus, and poor wound healing have been found in 9% of patients (Fazio, 1987).

Recurrence of Abscess or Fistula

After incision and drainage of ischioanal and intersphincteric abscesses, abscesses or fistulas recur in 89% of patients (Vasilevsky and Gordon, 1984). Recurrence appears in patients with a history of abscess drainage due to destruction of the natural barriers to infection, missed infection in adjacent anatomic spaces, the presence of an undiagnosed fistula or abscess at initial abscess drainage, and failure to completely drain the abscess (Abcarian, 1982; Schouten and van Vroonhoven, 1991). Last, failure to detect a primary opening of the fistula at the time of primary fistulotomy and abscess drainage or to recognize lateral or upward extensions of a fistula may result in persistence of the infection and recurrence. Other causes, including the inability to locate the primary opening of the fistula, spontaneous closure of the primary opening, microscopic opening of the tract, and the presence of secondary tracts (which can be easily missed) accounted for early recurrence in 20%. Recurrence rates after fistulotomy range from 0% to 18% (Vasilevsky and Gordon, 1985; Sangwan et al., 1994). Epithelialization of the fistula tract from internal or external openings rather than chronic infection of an anal gland may lead to recurrence (Lunniss et al., 1995).

Patients usually complain of intermittent purulent or bloody discharge with pain. There is an episode of acute anorectal sepsis that usually settles spontaneously with or without antibiotics. The passage of flatus or feces through the external opening is suggestive of a rectal, rather than an anal, fistula. Good postoperative care can also reduce recurrence rates by avoiding bridging and pocketing of the wound (Soew-Choen and Phillips, 1991).

Recurrence rates of fistula after treatment with seton ranged from 0% to 29% (Pearl et al., 1993), while recurrence under management with flap was 40% (Mizrahi et al., 2002). This high recurrence rate can be minimized by using full-thickness rectal wall to avoid necrosis or retraction of the flap (Lewis and Bartolo, 1990). Recurrence of abscess and fistula can be prevented by using antibiotics (Sözener et al., 2011; Hasan, 2017) and treating the causative disease, for example, Crohn's disease. In addition, performing a fistulotomy at the same time the abscess is drained, and sending drained pus for culture and sensitivity may reduce the rate of recurrence (Benjelloun et al., 2013).

Incontinence
The incidence of anal incontinence was 39% (Schouten and van Vroonhoven, 1991). Incontinence may result after treatment of the abscess by incision and drainage either from iatrogenic damage to the sphincter or inappropriate wound care by the patient. Other causes are repeated drainage for recurrence, fistulectomy, necrotizing fasciitis, and Crohn's disease. Continence may be compromised if the superficial external sphincter is accidentally divided during drainage of a perianal or deep postanal abscess in a patient with preoperative borderline continence. Drainage of a supralevator abscess may lead to incontinence if the puborectalis muscle is incorrectly divided (Seow-Choen and Nicholls, 1992). In addition, incontinence may result from prolonged packing of a drained abscess by preventing the development of granulation tissue and promoting the formation of excess scar tissue (Mazier, 1971). Treatment of fistula by primary fistulotomy may result in division of external sphincter muscle in acutely inflamed tissue.

Treatment of anterior fistulas in elderly patients, especially females, has an increased incidence of incontinence (as mentioned previously regarding management).

Scarring and Stenosis
This may occur due to large incision of the abscess and repeated drainage procedures of due to multiple recurrences of the abscess and fistulas. Anal stenosis may result from healing of the anal canal by scar contracture (Wexner et al., 1996).

Anal Mucosal Prolapsed
This is caused by extensive division of sphincter muscle and can be treated by band ligation, sclerosis, or excision (Fazio, 1987).

Chronic Arthritis of the Hip Joint

Chronic arthritis of the hip joint is a very rare complication that occurred in a healthy, young, male athlete as a complication of an inadequately treated anal fistula. On examination, he was found to have a tender right hip joint with severe restriction of movement and a partially drained right ischiorectal abscess. X-ray and MRI of the hip joint revealed chronic arthritis of the right hip joint, which was communicating with a complex fistula-in-ano. Management involved diversion sigmoid colostomy and right ischiorectal abscess drainage, along with appropriate antibiotics and hip joint replacement later on (Raghunath et al., 2014).

REFERENCES

Abcarian, H., 1982. Surgical management of recurrent anorectal abscess. Contemp. Surg. 21, 85–91.

Benjelloun, E.B., Jarrar, A., El Rhazi, K., Souiki, T., Ousadden, A., Ait Taleb, K., 2013. Acute abscess with fistula: long-term results justify drainage and fistulotomy. Updat. Surg. 65, 207–211.

Fazio, V.W., 1987. Complex anal fistulae. Gastroenterol. Clin. N. Am. 16, 93–114.

Hasan, R.M., 2017. Postoperative empirical antibiotic use for uncomplicated perianal abscess and fistula. Ann. Colorectal Res. 5, e40795.

Lewis, P., Bartolo, D.C.C., 1990. Treatment of trans-sphincteric fistulae by full thickness anorectal advancement flaps. Br. J. Surg. 77, 1187–1189.

Lunniss, P.J., Sheffield, J.P., Talbot, I.C., et al., 1995. Persistence of idiopathic anal fistula may be related to epithelialization. Br. J. Surg. 82, 32–33.

Mazier, W.P., 1971. The treatment and care of anal fistulas: a study of 1000 patients. Dis. Colon Rectum 14, 134–144.

Mizrahi, N., Wexner, S.D., Zmora, O., et al., 2002. Endorectal advancement flap: are there predictors of failure? Dis. Colon Rectum 45, 1616–1621.

Pearl, R.K., Andrews, J.R., Orsay, C.P., et al., 1993. Role of the seton in the management of anorectal fistulas. Dis. Colon Rectum 36, 573–579.

Raghunath, R., Varghese, G., Simon, B., 2014. Chronic arthritis of the hip joint: an unusual complication of an inadequately treated fistula-in-ano. BMJ Case Rep. 20. pii: bcr2014207086.

Sangwan, Y.P., Rosen, L., Riether, R.D., et al., 1994. Is simple fistula-inano simple? Dis. Colon Rectum 37, 885–889.

Schouten, W.R., van Vroonhoven, T.M.J.V., 1991. Treatment of anorectal abscesses with or without primary fistulectomy: results of a prospective randomized trial. Dis. Colon Rectum 34, 60–63.

Seow-Choen, F., Nicholls, R.J., 1992. Anal fistula. Br. J. Surg. 79, 197–205.

Soew-Choen, F., Phillips, R.K.S., 1991. Insights gained from the management of problematical anal fistulae at St. Mark's hospital, 1984–88. Br. J. Surg. 78, 539–541.

Sözener, U., Gedik, E., Kessaf Aslar, A., Ergun, H., Halil Elhan, A., Memikoğlu, O., Bulent Erkek, A., Ayhan Kuzu, M., 2011. Does adjuvant antibiotic treatment after drainage of anorectal abscess prevent development of anal fistulas? A randomized, placebo-controlled, double-blind, multicenter study. Dis. Colon Rectum 54, 923–929.

Vasilevsky, C.A., Gordon, P.H., 1984. The incidence of recurrent abscesses or fistula-in-ano following anorectal suppuration. Dis. Colon Rectum 27, 126–130.

Vasilevsky, C.A., Gordon, P.H., 1985. Results of treatment of fistula-in-ano. Dis. Colon Rectum 28, 225–231.

Wexner, S.D., Rosen, L., Roberts, P.L., et al., 1996. Practice parameters for treatment of fistula-in-ano: supporting documentation. Dis. Colon Rectum 39, 1363–1372.

FURTHER READING

Vasilevsky, C.A., Gordon, P.H., 2007. Benign anorectal sepsis. In: The ASCRS Text Book of Colon and Rectal Surgery. Vol. 13, p. 192.

INDEX

Note: Page numbers followed by *f* indicate figures.

Printed in the United States
By Bookmasters